Daily *warm-ups*

ALGEBRA II

Hope Martin

WALCH **PUBLISHING**

1   2   3   4   5   6   7   8   9   10

ISBN 0-8251-5084-1

Copyright © 2004

Walch Publishing

P. O. Box 658 • Portland, Maine 04104-0658

walch.com

Printed in the United States of America

Table of Contents

To the Teacher iv

States of the Union 1–34

Famous Firsts...35–68

Discoveries, Inventions, and Notable Accomplishments 69–102

Happy Birthday to You..103–147

Historical Highlights 148–180

Answer Key..181

Glossary of Terms 186

The *Daily Warm-Ups* series is a wonderful way to turn extra classroom minutes into valuable learning time. The 180 quick activities—one for each day of the school year—review, practice, and teach algebra. These daily activities may be used at the very beginning of class to get students into learning mode, near the end of class to make good educational use of that transitional time, in the middle of class to shift gears between lessons—or whenever else you have minutes that now go unused. In addition to providing students with fascinating information, they are a natural path to other classroom activities involving critical thinking.

While all of the historical events in the *Pre-Algebra, Algebra,* and *Algebra II Daily Warm-Ups* are the same, the clues given to solve the puzzles correlate with the mathematics skills and concepts applicable to each level of learning. This allows you as the teacher to facilitate differentiation of math instruction by easily providing multiple approaches to content. At the same time, each student is given the opportunity to solve the problem and his or her ability level.

In the middle school, *Daily Warm-Ups: Pre-Algebra* can effectively be used in combination with *Daily Warm-Ups: Algebra.* At the high school level, any two or all three can be used in combination to better meet the learning needs of students.

Daily Warm-Ups are easy-to-use reproducibles—simply photocopy the day's activity and distribute it. Or make a transparency of the activity and project it on the board. You may want to use the activities for extra-credit points or as a check on critical-thinking and problem-solving skills.

However you choose to use them, *Daily Warm-Ups* are a convenient and useful supplement to your regular lesson plans. Make every minute of your class time count!

States of the Union

The growth of the United States took place over the course of 172 years. Delaware was the first of the colonies to join the Union (December 7, 1787) and Hawaii became the 50th state (August 21, 1959).

The mathematics puzzles in this chapter help celebrate the nation we are today. When students solve the mathematical clues, they learn the year that each state became a part of the United States of America. Interdisciplinary discussions can encourage students to explore the connections between the dates colonies and territories became states, colonial times, pioneer exploration, and the growth of the western part of the United States. And, when students investigate mathematics within the framework of history, they view math in context, and it begins to make sense.

Delaware, Pennsylvania, and New Jersey

were the first three colonies to enter the Union. They all became states in the month of December of the same year. Solve this puzzle to learn the year.

- My hundreds and units digits are the same number, and each is the value of x when this equation is solved: $2x^2 - 11 = 87$.

- My tens digit is equal to the value of x: $-5x = -40$.

- The sum of all of my digits is equal to $3^3 - 2^2$.

What year am I?

———————   ———————   ———————   ———————
Thousands   Hundreds   Tens   Units

Georgia, Massachusetts, Connecticut,

New Hampshire, South Carolina, Virginia, New York, and Maryland all became states in the same year. They became the 4th through the 11th states of our United States. Solve this puzzle to learn the year.

- My tens and units digits are the same number. Learn them by evaluating this expression: $7 - 4 \div 12 + 8 \div 6$.

- My hundreds digit is equal to the value of y in this equation: $5(y - 2) = 2y + 11$.

- The sum of all of my digits is equal to $\sqrt{625} - 1$.

What year am I?

2

_____   _____   _____   _____

Thousands   Hundreds   Tens   Units

North Carolina became the 12th colony to join the Union. It became a state on November 21 of this year. To learn the year, just solve this puzzle.

- The two-digit number formed by my hundreds and tens digits is equal to the value of x: $3x + 40 = 2x + 118$.

- My units digit is one greater than my tens digit.

- The sum of all of my digits is equal to $\sqrt{625}$.

What year am I?

Thousands	Hundreds	Tens	Units

© 2004 Walch Publishing

Rhode Island became the last of the 13 original colonies when it became a state on May 29th. To find the year that Rhode Island became a state, just solve this puzzle.

- The two-digit number formed by my thousands and hundreds digits is equal to the value of x: $4(3x - 1) = 11x + 13$.

- My tens digit is equal to the value of z in this equation: $4z - 3(z + 2) = \frac{1}{3}z$.

- The sum of all of my digits is the 7th prime number.

What year am I?

_____   _____   _____   _____
Thousands   Hundreds   Tens   Units

On March 4 of this year, Vermont became the 14th state to join the Union. To learn the year that Vermont became a state, just solve this puzzle.

- My hundreds and tens digits are consecutive odd integers, the sum of which is 16.

- My units digit is equal to the solution to this inequality: $4 - 5y > -1$.

- The sum of all of my digits is equal to $2\sqrt{81}$.

What year am I?

Thousands	Hundreds	Tens	Units
_____	_____	_____	_____

5

Kentucky became the 15th state of the Union when it became a state on June 1 of this year. To learn the year Kentucky gained statehood, just solve this puzzle.

- Examine this equation to discover my hundreds and tens digits: $y = 7x + 9$. My hundreds digit is the slope, and my tens digit is the y-intercept.

- My units digit is equal to the slope of the line containing these two points: $(2, 8)$, $(0, 4)$.

- The sum of all of my digits is the value of x in this equation: $\sqrt{100} + \sqrt{81} = x$.

What year am I?

6

_____ _____ _____ _____
Thousands Hundreds Tens Units

Tennessee became the 16th state of the United States when it joined the union on June 1 of this year. To learn the year Tennessee became a state, just solve this puzzle.

• Study these equations. My hundreds digit is the value of x, my tens digit is the value of y:

$$\begin{cases} x + y = 16 \\ y = x + 2 \end{cases}$$

• My units digit is a perfect number.

• The sum of all of my digits is $2^5 - 3^2$.

What year am I?

_____   _____   _____   _____
Thousands   Hundreds   Tens   Units

7

On March 1 of this year, Ohio became the 17th state of the United States. To learn the year, just solve this puzzle.

- Study these equations. My hundreds digit is equal to the value of x, my units digit is equal to the value of y.

$$\begin{cases} 2x = 5y + 1 \\ x - y = 5 \end{cases}$$

- My tens digit is equal to $g(-3)$ in this function: $g(x) = x^2 - x - 12$.

- The sum of all of my digits is equal to $\sqrt{144}$.

What year am I?

8

_____   _____   _____   _____
Thousands   Hundreds   Tens   Units

Louisiana became the 18th state of the Union on April 30 of this year. To learn the year that Louisiana became a state, just solve this puzzle.

- The two-digit number formed by tens and units digits could be the value of x if the perimeter of this rectangle equals 138 units.

$2x$ ⎕ $4x - 3$

- My hundreds digit is four times my units digit.

- The sum of my digits is two more than the sum of the first four counting numbers.

What year am I?

_____   _____   _____   _____
Thousands   Hundreds   Tens   Units

9

On December 11 of this year, Indiana became the 19th state to join the Union. Solve this puzzle to learn the year.

- The two-digit number formed by my tens and units digit is the solution to this problem:

  "If you subtract 15 from this number, multiply the difference by 4, find the positive square root of the product, and subtract 2, the difference is 0. What is this two-digit number?"

- My hundreds digit is the value of y in this equation: $2y + y - 3 = 21$.

- One hundred fifty percent of the sum of all of my digits is 24.

What year am I?

10

| _____ | _____ | _____ | _____ |
| Thousands | Hundreds | Tens | Units |

Mississippi became a state on December 10 of this year. It was the 20th state to join the Union. Solve this puzzle to learn the year.

- The two-digit number formed by my thousands and hundreds digits is equal to the value of n in this equation: $-\frac{1}{2}n = -9$.
- My units digit is equal to the value of y: $12y - 8(2y - 1) = -20$.
- The sum of all of my digits is $\sqrt{289}$.

What year am I?

_____   _____   _____   _____

   Thousands   Hundreds   Tens   Units

11

On December 3 of this year, Illinois became the 21st state to join the Union. To learn the year that Illinois became a state, just solve this puzzle.

- The two-digit number formed by my tens and units digits is the value of x in this proportion: $\dfrac{4}{6+x} = \dfrac{2}{x-6}$

- My hundreds digit is the y-intercept in the equation $2x + y = 8$.

- The sum of all of my digits is $\sqrt[3]{5832}$.

What year am I?

12

$\overline{}$ $\overline{}$ $\overline{}$ $\overline{}$

Thousands   Hundreds   Tens   Units

© 2004   Walch Publishing

Alabama became the 22nd state of the Union on December 14 of this year. To learn the year that Alabama became a state, just solve this puzzle.

- Wendy is 3 years older than Shawn, Lynne is 1 year younger than Shawn. In 5 years, the sum of their ages will be 65 years. Find Wendy's age, and you will know the two-digit number formed by my tens and units digits.

- The two-digit number formed by my thousands and hundreds digit is equal to the value for y when $x = 13$ in this linear equation: $y = 2x - 8$.

- The sum of all of my digits is the prime number that satisfies this inequality: $17 \leq n \leq 23$.

What year am I?

| _____ | _____ | _____ | _____ |
| Thousands | Hundreds | Tens | Units |

13

On March 15 of this year, Maine became the 23rd state of the United States. To learn the year this happened, just solve this puzzle.

- My hundreds digit is the slope of the line containing the two points whose coordinates are $(1, 1)$ and $(3, 17)$.

- The two-digit number formed by my tens and units digits is $f(3)$ for the function $f(x) = x^2 + 3x + 2$.

- The sum of all of my digits is the value of m in this equation: $4m + 6 = 5m - 5$.

What year am I?

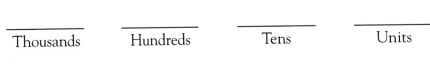

| Thousands | Hundreds | Tens | Units |

14

Missouri became the 24th state on August 10 of this year. To learn when Missouri became a state, just solve this puzzle.

- My hundreds digit is equal to $2(\sqrt[3]{64})$.

- The two-digit number formed by my tens and units digits is the solution to this problem: "A car uses 7 gallons of gasoline to travel 63 miles; how much gasoline would it take to travel 189 miles?"

- On a certain road map, a map distance of 1 cm represents an actual distance of 30 km. The sum of all of my digits is equal to the map distance that represents an actual distance of 360 km.

What year am I?

| _____ | _____ | _____ | _____ |
| Thousands | Hundreds | Tens | Units |

15

On June 10 of this year, Arkansas became the 25th state of the United States. Solve this puzzle to learn the year.

- The two-digit number formed by my tens and units digits is the solution to the composite function $f[g(4)]$ when $f(x) = x^2$ and $g(x) = x + 2$.

- My hundreds digit is the y-intercept for the line formed by the ordered pairs $(2, 14)$ and $(-4, -4)$.

- The sum of all of my digits is the perimeter of this rectangle when $x = 2$:

$$4x - 3$$

$2x$ ⬚

What year am I?

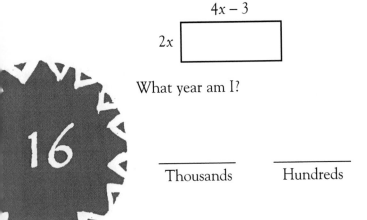

16

_____   _____   _____   _____
   Thousands   Hundreds   Tens   Units

Michigan became the 26th state of the United States on January 26 of this year. Solve this puzzle to learn the year.

- Michelle's mother is 1 year older than twice Michelle's age. In 5 years, the sum of their ages will be 65. The two-digit number formed by my thousands and hundreds digits is the same as Michelle's age; the two-digit number formed by my tens and units digits is equal to her mother's age.

- The sum of my digits is the positive solution to this quadratic equation: $x^2 - 17x - 38 = 0$.

What year am I?

Thousands	Hundreds	Tens	Units

17

In this year, both Florida and Texas became the 27th and 28th states of the United States. Solve this puzzle to learn the year.

- The two-digit number formed by my tens and units digits is the value of this expression when $x = 4$: $(3x^2 - 5x) + (5x - 3)$.

- My hundreds digit is equal to the degree of the polynomial $-6s^2t^4u^2$.

- The sum of all of my digits is the GCF of 54 and 90.

What year am I?

Thousands	Hundreds	Tens	Units
_____	_____	_____	_____

18

On December 28 of this year, Iowa became the 29th state of the United States. Solve this puzzle to learn the year.

- The two-digit number formed by my tens and units digit is equal to $\{f \circ g\}(8)$ if $f(x) = 2x + 2$ and $g(x) = 3x - 2$.

- If $g(x) = 3x + 7$, my hundreds digit is equal to $g(\frac{1}{3})$.

- The sum of all of my digits is the prime number (p) that satisfies this inequality: $17 \leq p \leq 23$.

What year am I?

Thousands	Hundreds	Tens	Units

19

Wisconsin became the 30th state of the United States on May 29 of this year. To learn the year Wisconsin became a state, just solve this puzzle.

- Simplify this complex fraction to find my units and hundreds digits: $\dfrac{\frac{4}{5}}{\frac{1}{15} + \frac{1}{30}}$.

- My tens digit is the value of a: $\dfrac{a + 5}{3} = \dfrac{a + 2}{2}$.

- The sum of all of my digits is equal to $|3(-5 + (-2))|$.

    What year am I?

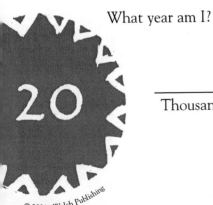

20

_____   _____   _____   _____
Thousands   Hundreds   Tens   Units

21

On September 9 of this year, California became the 31st state of the United States. Solve this puzzle to learn the year.

- The two-digit number formed by my thousands and hundreds digits is equal to the value of y: $\dfrac{1}{3} = \dfrac{0.5y - 1}{y + 6}$.

- My tens digit is the slope:

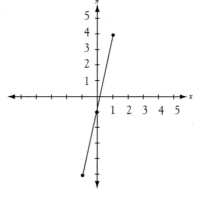

- The sum of all of my digits is equal to $\sqrt{225 - 29}$.

What year am I?

© 2004 Walch Publishing

Thousands ___ Hundreds ___ Tens ___ Units ___

On May 11 of this year, Minnesota became the 32nd state of the United States. Solve this puzzle to learn the year.

- When you solve this system of equations, you will discover the value of all of my digits. My hundreds and units digits are equal to the value of x, my tens digit the value of y, my thousands digit the value of z:

$$\begin{cases} x = y + 3z \\ x + y + z = 21 \\ y - z = x - 4 \end{cases}$$

- The sum of all of my digits is a multiple of 11.

What year am I?

22

$\overline{\hspace{3em}}$ $\overline{\hspace{3em}}$ $\overline{\hspace{3em}}$ $\overline{\hspace{3em}}$
Thousands   Hundreds   Tens   Units

Oregon became the 33rd state of the United States on February 14 of this year. Solve this puzzle to learn the year.

- When you solve this system of equations, you will discover the value of all of my digits. My units digit is equal to the value of x, my tens digit to the value of y, my hundreds digit to the value of z:

$$\begin{cases} x + y + z = 22 \\ x = z + 1 \\ z + y = x + 4 \end{cases}$$

- The sum of my digits is equal to $5 + 2^3 + \sqrt{16} + 3 \times 2$.

What year am I?

Thousands	Hundreds	Tens	Units

23

© 2004 Walch Publishing

Kansas became the 34th state of the United States on January 29 of this year. To learn the year this event took place, just solve this puzzle.

- My hundreds digit is the determinant of this 2×2 matrix:

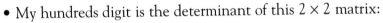

$$\begin{bmatrix} 3 & 2 \\ 2 & 4 \end{bmatrix}$$

- My tens digit is equal to the value of this expression:

$$[(-2)^3 + 2](\sqrt{9} - 6) \div 3.$$

- The sum of all of my digits is equal to the value of r in this problem: Marla rode her bicycle at an average rate (r mph) for 3.5 hours. She traveled a distance of 56 miles. What was her average speed?

What year am I?

24

_____   _____   _____   _____
Thousands   Hundreds   Tens   Units

West Virginia became the 35th state of the United States on June 20 of this year. To learn the year of this event, just solve this puzzle.

- To find the value of my hundreds and tens digits, solve this system of equations. My hundreds digit is equal to the value for y; my tens digit is equal to the value of x:

$$\begin{cases} \dfrac{1}{3}x + \dfrac{1}{2}y = 6 \\ 4x - 2y = 8 \end{cases}$$

- To find the value of my units digit, find the value of y (the value of x is given): $y = \dfrac{1}{3}x - 2$, $x = 15$.

- The sum of my digits is equal to $\dfrac{4^2 + 5(2) + \sqrt{100}}{2}$.

What year am I?

Thousands	Hundreds	Tens	Units
_____	_____	_____	_____

On October 31 of this year, Nevada became the 36th state of the United States. To learn the year this happened, just solve this puzzle.

- My units digit is the solution to this inequality:

  $-3x + 5 \leq -7$. My units digit $x \geq$.

- My hundreds digit is the slope, and my tens digit is the y-intercept of the line containing these two points: (1, 14) and (−3, −10).

- The amount of oxygen a person's body is able to use is about $3\frac{1}{2}$ liters per minute. The sum of my digits is equal to the liters of oxygen that can be used in $5\frac{3}{7}$ minutes.

What year am I?

26

| Thousands | Hundreds | Tens | Units |

Nebraska, our 37th state, joined the Union on March 1 of this year. To learn the year Nebraska became a state, just solve this problem.

- Solve this system of equations. My tens digit is the same as the value of x; my units digit is the same as the value of y:

$$\begin{cases} y = 2x - 5 \\ 2x + 3y = 33 \end{cases}$$

- The two-digit number formed by the thousands and hundreds digits is the determinant of this 2×2 matrix:

$$\begin{bmatrix} 5 & 4 \\ 3 & 6 \end{bmatrix}$$

- The sum of my digits is the value of this expression when $x = 2$ and $y = 1$:
$2x^2 + 2xy + 10y^2$.

What year am I?

27

---------   ---------   ---------   ---------

Thousands   Hundreds   Tens   Units

Colorado became the 38th state of the Union on August 1 of this year. To learn the year this happened, just solve this puzzle.

- The two-digit number formed by my thousands and hundreds digits is the value of x: $\sqrt{2x} = 6$.

- My tens digit is the value of r: $\sqrt{r^2} = 7$.

- The sum of all of my digits is equal to the value of y: $\sqrt[3]{y+5} = 3$.

What year am I?

Thousands	Hundreds	Tens	Units

28

In this year, four more states were added to the Union. North and South Dakota were admitted on the same day, November 2; Montana followed on November 8, and Washington was admitted 3 days later. Solve this puzzle to learn the year.

- Graph this system of equations, and find its solution. The *x*-coordinate is the same as my tens digit; the *y*-coordinate is the same as my units digit.

$$\begin{cases} y = x + 1 \\ x + y = 17 \end{cases}$$

- The two-digit number formed by my thousands and hundreds digits is equal to the value of *t*: $\sqrt{t+7} = 5$.

- The sum of all of my digits is the larger of two consecutive positive integers, the product of which is 650.

What year am I?

29

Thousands	Hundreds	Tens	Units

In July of this year, Idaho and Wyoming were admitted as the 43rd and 44th states of the Union. Idaho was admitted on July 3 and Wyoming on July 10. To learn the year, just solve this puzzle.

- The sum of my hundreds and tens digits is 17. The

  $$\sqrt[3]{\text{hundreds digit}} + 1 = \sqrt{\text{tens digit}}.$$

- Label the graph and graph this inequality. My units digit is its upper boundary: $\frac{x}{5} + 2 < 2$.

- The sum of all of my digits is equal to the value of w: $\dfrac{2}{w+4} = \dfrac{3}{w+15}$

  What year am I?

| _____ | _____ | _____ | _____ |
| Thousands | Hundreds | Tens | Units |

© 2004 Walch Publishing

Utah became the 45th state in January of this year. Utah was admitted to the Union on January 4. To learn the year, just solve this puzzle.

- My tens digit is the *y*-intercept of the line that contains the points (4, 11) and (–6, 6).

- My units digit is the value of *n*: $\sqrt[3]{n+21} = 3$.

- My hundreds digit is equal to *f*(2) for $f(x) = 2x^2$.

- Solve this problem to find the sum of all of my digits:
  "The sum of the digits is 6. When the digits of my sum are reversed, the resulting number is 18 greater than the original number."

What year am I?

Thousands	Hundreds	Tens	Units
_____	_____	_____	_____

31

Oklahoma became the 46th state when it was admitted to the Union on November 16 of this year. To learn the year, just solve this puzzle.

- My units digit is the GCF of 126 and 385.

- The sum of three consecutive integers is 273. Find the smallest of the three integers to learn the two-digit number formed by my hundreds and tens digits.

- The sum of all of my digits is the product that results when these factors are multiplied: $(4 + i)(4 - i)$.

What year am I?

32

Thousands	Hundreds	Tens	Units
_____	_____	_____	_____

In this year, New Mexico and Arizona became the 47th and 48th states admitted to the Union. New Mexico became a state on January 6 and Arizona on February 14. To learn the year, just solve this puzzle.

- The two-digit number formed by my tens and units digit is the solution to $\sqrt[3]{2x+3} = 3$.

- To find my thousands and hundreds digit, solve this system of equations; x is my thousands digit and y is my hundreds digit:

$$\begin{cases} y + 3x + 6 \\ x^2 + 9 = 10 \end{cases}$$

- The sum of all of my digits is the value of k ($y = a(x - h)^2 + k$) when the square of this quadratic is completed:
$y = x^2 + 8x - 3$.

What year am I?

_____   _____   _____   _____

Thousands   Hundreds   Tens   Units

33

The last two states to be admitted to the Union,

Alaska (49th) and Hawaii (50th) were both admitted in the same year. Alaska was admitted on December 3 and Hawaii on August 21. To learn the year, just solve this puzzle.

- The sum of three angles (x, y, and z) is 180°. The sum of angles x and y is equal to $z - 24°$. Angle y is 40° larger than angle x. The two-digit number formed by my tens and units digits is equal to the number of degrees in angle x; the two-digit number formed by my thousands and hundreds digits is equal to the number of degrees in angle y.

- The sum of all of my digits is equal to the number of pints of a 60% acid solution that must be mixed with 12 pints of a 15% acid solution to produce a 45% acid solution.

What year am I?

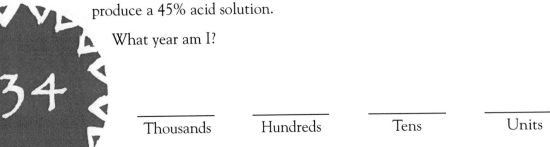

34

_____ _____ _____ _____
Thousands   Hundreds   Tens   Units

© 2004 Walch Publishing

Famous Firsts

We are all fascinated by "the first of anything" — The first person to climb a mountain, the first minority to be elected to office, even the first telephone operator! These puzzles take us back in time to contemplate those explorers, scientists, entrepreneurs, and others who have changed history by being the "first of something"! By solving these puzzles, students learn the year that the "famous first" occurred, while they communicate "mathematically."

On January 5 of this year, the first junior high school was opened in the United States. Solve this puzzle to learn the year.

- If you subtract 6 from the two-digit number formed by my tens and units digits, then multiply that difference by 4, then take the positive square root of the product and finally subtract 4, the result would be 0.

- My hundreds digit is equal to the value of n in this equation: $\frac{2}{3} n + 4 = n + 1$.

- If twice the sum of all my digits is increased by 3, the result is 8 less than 3 times my sum.

What year am I?

_____   _____   _____   _____
Thousands   Hundreds   Tens   Units

Nellie Tayloe Ross became the first woman to be elected a governor in the United States. She took the oath of office on January 5 of this year. The state was Wyoming.

- The sum of the second and third of three consecutive integers is 3 more than twice the first. The two-digit number formed by my tens and units digits is the smallest of these three consecutive integers.

- David is 12 years older than Samantha is now. In 5 years, he will be $1\frac{1}{2}$ times her age. The two-digit number formed by my thousands and hundreds digits is Samantha's age now.

- The sum of all of my digits is equal to the value of c in this equation: $-5.73c = -97.41$.

What year am I?

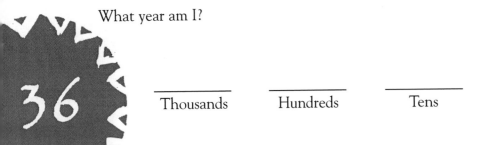

36

_____   _____   _____   _____

Thousands   Hundreds   Tens   Units

On January 7 of this year, the first boat to cross "the path between the seas" (the crane boat *Alex La Valley*) sailed from the Atlantic Ocean to the Pacific Ocean when the Panama Canal opened. Find the year this happened by solving this puzzle.

- A parking lot charges \$1.25 for the first hour and \$0.75 for each additional hour or fraction of an hour. Margo has only \$11.00 to spend on parking. The two-digit number formed by my tens and units digit is equal to the maximum number of hours Margo can park and stay within her budget.

- My hundreds digit is equal to one more than twice my units digit.

- The length of a rectangle is one less than twice its width. Its perimeter is 46. The sum of all of my digits is equal to its length.

What year am I?

37

_____   _____   _____   _____
Thousands   Hundreds   Tens   Units

The first woman U.S. Senator, Hattie Caraway, was elected from the state of Arkansas in this year. Solve the puzzle to learn the year.

- The two-digit number formed by my tens and units digits is equal to the value of this expression when $x = 3$ and $y = 2$: $2x^2 + y^3 + 3$!

- My hundreds digit is equal to z in this equation: $\dfrac{4z + 8}{2} = 22$.

- The sum of all of my digits is equal to $5! \div 2^3$.

What year am I?

| Thousands | Hundreds | Tens | Units |

38

On January 13 of this year, Robert C. Weaver became the first African-American cabinet member when President Lyndon Johnson appointed him Secretary of Housing and Urban Development. Learn the year by solving this puzzle.

- My tens and units digits are the same number. For the function $g(x) = \dfrac{x4 + 8x - 2}{x2 + 3x - 5}$, solve for $g(2)$ to find these two numbers.

- The two-digit number formed by my thousands and hundreds digit is the larger of two consecutive positive integers, the product of which is 342.

- The sum of all of my digits is equal to the $5\sqrt{16} + 2$.

What year am I?

_____ _____ _____ _____
Thousands   Hundreds   Tens   Units

39

On January 21 of this year, the first law requiring that drivers of automobiles have licenses went into effect. To learn the year, solve this puzzle.

- My hundreds digit is the unknown leg of this right triangle:

x

- The two-digit number formed by my tens and units digits is the determinant of this matrix:

$$\begin{bmatrix} 2 & -1 \\ 29 & 4 \end{bmatrix}$$

- The sum of all of my digits is the sum of the first four triangular numbers.

    What year am I?

Thousands	Hundreds	Tens	Units

In this year, Dr. Elizabeth Blackwell became the first woman to receive an M.D. degree. She attended the Medical Institute of Geneva, New York. Solve this puzzle to find the year.

- The two-digit number formed by my tens and units digits is the sum of the first 7 odd numbers. The two-digit number formed by my hundreds and tens digits is the GCF of 54 and 126.

- The sum of all of my digits is equal to $g(5)$ for the function $g(x) = x^2 - 3$.

What year am I?

_____   _____   _____   _____
Thousands   Hundreds   Tens   Units

41

© 2004 Walch Publishing

On January 28 of this year, Louis Brandeis became the first American Jew to be appointed to the Supreme Court. Solve this puzzle to learn the year.

- The two-digit number formed by my hundreds and tens digits is the value of this expression: $2^3(3^2 - \sqrt[3]{8}) + (7 \times 5)$.

- My units digit is equal to $\dfrac{\sqrt{48x^2}}{\sqrt{8x^2}}$.

- The sum of all of my digits is $(1.7 \times 10^2) \div (1.0 \times 10^1)$.

What year am I?

42

Thousands   Hundreds   Tens   Units

On January 31 of this year, the first daytime soap opera, *These Are My Children*, was broadcast from the NBC studios in Chicago. Solve this puzzle to learn the year.

- The two-digit number formed by my tens and units digits is the value of n: $\log_7 n = 2$.

- My hundreds digit is the slope of the equation $-9x + y = 2$.

- The sum of all of my digits is equal to the y-intercept of the line $-7x + y = 23$.

What year am I?

_____   _____   _____   _____
Thousands   Hundreds   Tens   Units

43

Maggie Walker became the first African-American woman to establish and manage a bank in this year. Solve this puzzle to learn the date.

- Solve this system of equations. My hundreds digit is equal to the value of x and my units digit to the value of y:

$$\begin{cases} 2x - 5y = 3 \\ -x + 2y = -3 \end{cases}$$

- My thousands digit is equal to the value of x.

$$\begin{bmatrix} 2x & 7 \\ 4 & -6 \end{bmatrix} + \begin{bmatrix} 4 & -2 \\ -2 & -1 \end{bmatrix} = \begin{bmatrix} 6 & 5 \\ 2 & -7 \end{bmatrix}$$

- The sum of all of my digits is equal to the number between $\sqrt{144}$ and $\sqrt{196}$.

What year am I?

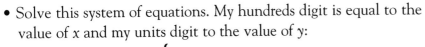
Daily Warm-Ups: Algebra II

44

_____   _____   _____   _____
Thousands   Hundreds   Tens   Units

On February 10 of this year, Western Union delivered the first "singing telegram." Solve the puzzle to learn the year.

- The two-digit number formed by my thousands and hundreds digits is the determinant of this matrix:

$$\begin{bmatrix} 4 & -7 \\ 1 & 3 \end{bmatrix}$$

- My tens digit and units digit are the same number; they are equal to the positive solution to this equation: $\frac{2}{3}x^2 = 6$.

- The sum of all of my digits is equal to the solution to this expression: $3x^2 - 5x + 14$ when $x = 2$.

What year am I?

_____	_____	_____	_____
Thousands	Hundreds	Tens	Units

45

On February 13 of this year, the first magazine was published in the United States. Called *American Magazine*, it was published by Andrew Bradford. Solve the puzzle to find the year.

- My hundreds and units numbers are the same. They are the *x*-intercept in this equation: $3x + 7y = 21$.

- If $f(x) = 2x + 1$ and $g(x) = x - 1$, then my tens digit is equal to $g[f(x)]$ for $x = 2$.

- The sum of all of my digits is equal to the number of degrees in the complement of a 71° angle.

What year am I?

46

_____   _____   _____   _____

Thousands   Hundreds   Tens   Units

Daily Warm-Ups: Algebra II

© 2004 Walch Publishing

On February 20 of this year, astronaut John Glenn became the first American to orbit Earth. He circled the globe three times before landing.

- The length of a rectangle is 1.5 times its width. Its area is 54 square units. The length of the rectangle is my hundreds digit; the width is my tens digit.

- My units digit is the number of diagonals in a square.

- On Earth, an object falls at the rate of 32 ft/sec². The sum of all of my digits is equal to the distance an object would fall in $\frac{9}{16}$ sec.

What year am I?

————————— ————————— ————————— —————————
Thousands   Hundreds   Tens   Units

47

© 2004 Walch Publishing

On March 2 of this year, the first school for the blind in America was established in Massachusetts. Solve this puzzle to learn the year.

- My thousands digit is equal to the axis of symmetry of this graph:

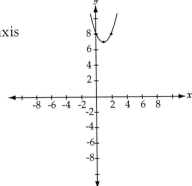

- Charlie is 4 times as old as his sister, Josie. In 5 years, the sum of their ages will be 20. My tens digit is equal to Josie's age; my hundreds digit is equal to Charlie's age.

- My units digit is the slope of the line that contains the coordinates (3,10) and (2,1).

- The sum of all of my digits is equal to the number of different pizzas with 3 cheeses that can be made if 6 different cheeses are available.

What year am I?

48

© 2004 Walch Publishing

| _____ | _____ | _____ | _____ |
| Thousands | Hundreds | Tens | Units |

Charles Brady King drove the first automobile in Detroit on March 6 of this year. Solve this puzzle to learn the year.

- The two-digit number formed by my tens and units digits is the score that Marsha needs to get on her fourth exam to have a 90 average. The scores on her first three were 85, 86, and 93.

- My hundreds digit is the y-intercept of this line: $-x + y = 8$.

- The sum of all of my digits is equal to 4!.

What year am I?

_____	_____	_____	_____
Thousands	Hundreds	Tens	Units

49

On March 12 of this year, the first parachute jump from an airplane in the United States was made. (What do you think the statement, "Minds are like parachutes, they only function when they're open" means?)

- The sum of two numbers is 11; their difference is 7. My hundreds digit is equal to x, and my units digit is equal to y.

- My tens digit is equal to the slope of this graph:

- The sum of all of my digits is the GCF of 65 and 78.

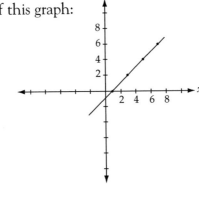

50

What year am I?

_____   _____   _____   _____
Thousands   Hundreds   Tens   Units

On March 16 of this year, the first black newspaper in the United States was published in New York City. It was called *Freedom's Journal*. The day is now celebrated as Black Press Day in New York.

- Find the solution to this system of equations:

$$\begin{cases} 2x + y = 12 \\ y = 4x \end{cases}$$

My tens digit is the same as the *x*-coordinate, my hundreds digit is the same as the *y*-coordinate.

- My units digit is the determinant of this matrix:

$$\begin{bmatrix} 3 & 4 \\ 2 & 5 \end{bmatrix}$$

- The sum of all of my digits is the middle number of three consecutive integers such that the sum of the second and third is 20 more than the first.

What year am I?

51

_____   _____   _____   _____

Thousands   Hundreds   Tens   Units

© 2004 Walch Publishing

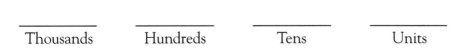

On March 19 of this year, the swallows first returned to the old mission of San Juan Capistrano, California. To learn the year, solve this puzzle.

- Each of the two congruent angles of an isosceles triangle is 20° more than twice the third angle. The two-digit number formed by my tens and units digits is the same as the number of degrees in each congruent angle.

- My hundreds digit is the inverse of the slope of the graph:

- The sum of all of my digits is the same as the sum of the first six counting numbers.

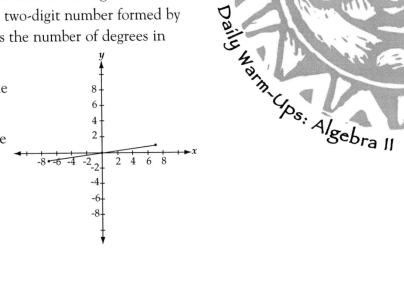

What year am I?

52

_____   _____   _____   _____
   Thousands   Hundreds   Tens   Units

On March 22 of this year, the first women's collegiate basketball game was played at Smith College in Northampton, Massachusetts. Sandra Berenson, the "Mother of Women's Basketball," supervised the game.

- Solve this system of equations, and find the value of x, y, and z. My hundreds digit is the value of x, my tens digit is the value of y, and my units digit is the value of z.

$$\begin{cases} x + y + z = 20 \\ 3z - 1 = x \\ x + z = y + 2 \end{cases}$$

- The sum of all of my digits is equal to the value of x in this equation: $\sqrt{36} - x = -15$.

What year am I?

_____	_____	_____	_____
Thousands	Hundreds	Tens	Units

53

Henry "Hank" Aaron hit the 715th home run of his career and broke Babe Ruth's record on April 8 of this year. He finished his career with a total of 755 home runs. Solve this puzzle to learn the year.

- Solve this system of equations:

$$\begin{cases} y = x - 3 \\ x + 2y = 15 \end{cases}$$

  The x-coordinate of its solution is the same as my tens digit; the y-coordinate is the same as my units digit.

- Two angles are complementary. One of the angles is 5° less than four times the measure of the other. The two-digit number formed by my thousands and hundreds digits is equal to the number of degrees in the smaller angle.

- The sum of all of my digits is equal to $\sqrt{25 \cdot 16} + 1$.

  What year am I?

54

Thousands	Hundreds	Tens	Units

On April 10 of this year, the Brooklyn Dodgers recruited Jackie Robinson. He was the first African-American to play on a major-league baseball club. Solve this puzzle to learn the year.

- The two-digit number formed by my thousands and hundreds digits is equal to the value of x in this angle:

$(x + 3)$

$68°$

- The number of chaperones required for a field trip varies in direct proportion to the number of students on the trip. If 3 chaperones are needed for 36 students, how many will be needed for 84 students? (This number is the same as my units digit.)

- The sum of all of my digits is equal to the total number of spots on one die.

What year am I?

55

Thousands	Hundreds	Tens	Units

On April 19 of this year, the Battle of Lexington began the Revolutionary War. The first gunfire was called "the shot heard round the world." To learn the year, solve this puzzle.

- My hundreds and tens digits are the same number. They are the slope of the line shown in this graph:

- My units digit is the solution of this function: $f(x) = \frac{x}{2} + 3$ for $f(4)$.

- The sum of all of my digits is equal to the number of two-letter combinations that can be made using the vowels A, E, I, O, U without repeating any letter.

What year am I?

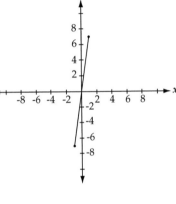

56

_____ _____ _____ _____
Thousands   Hundreds   Tens   Units

On April 25 of this year, the first Seeing Eye dog was presented in America. It has been found that purebred female German shepherds make the most effective Seeing Eye dogs. Solve this puzzle to find the year.

- My tens digit is the slope of the line that contains the coordinates (1, 5) and (−3, −3).

- My units digit is the y-intercept of the same line.

- The two-digit number formed by my thousands and hundreds digits is the solution of the function $f(x) = 3x + 4$ for $f(5)$.

- The sum of my digits is the same as the 5th triangular number.

What year am I?

| _____ | _____ | _____ | _____ |
| Thousands | Hundreds | Tens | Units |

57

Gwendolyn Brooks became the first African-American woman to win the Pulitzer Prize on May 5 of this year. Solve this puzzle to learn the year.

- The two-digit number formed by my tens and units digits is equal to the area of the triangle shown:

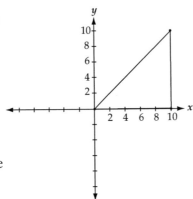

- My hundreds digit is equal to the solution of the function $f(g) = \frac{2}{3}g - 3$ when $g = 18$.

- The sum of all of my digits is equal to the sum of the first five counting numbers.

What year am I?

Daily Warm-Ups: Algebra II

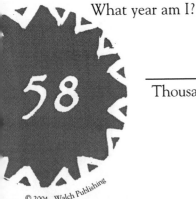

58

_____ _____ _____ _____
Thousands   Hundreds   Tens   Units

Hernando de Soto became the first European to reach the Mississippi River on May 8 of this year. Solve this puzzle to learn the year.

- Jill is 3 years younger than Joe, and Jackie is 1 year older than Joe. In 5 years, the sum of Jill's and Joe's ages will be 5 greater than Jackie's age. Jill's age now is the same as my units digit, Joe's age now is the same as my tens digit, and Jackie's age now is the same as my hundreds digit.

- The sum of all of my digits is equal to the determinant of this matrix:

$$\begin{bmatrix} 8 & 1 \\ 5 & 2 \end{bmatrix}$$

What year am I?

Thousands	Hundreds	Tens	Units

59

© 2004 Walch Publishing

The first ocean-to-ocean railroad was

completed on May 10 of this year. The Union Pacific and Central Pacific railways were linked at Promontory Point, Utah, by a golden spike. Learn the year by solving this puzzle.

- My hundreds digit is two greater than my tens digit; my units digit is $1\frac{1}{2}$ times greater than my tens digit. Their sum is 23.

- The sum of all of my digits is equal to the number of ways the letters "**H E L P**" can be arranged to form four-letter combinations, if letters cannot be repeated.

What year am I?

60

$\overline{\qquad}$ $\overline{\qquad}$ $\overline{\qquad}$ $\overline{\qquad}$
Thousands   Hundreds   Tens   Units

On May 10 of this year, the first planetarium in the United States opened in the city of Chicago. It is called the Adler Planetarium. Solve this puzzle to learn the year.

- The two-digit number formed by my tens and units digits is equal to $2^0 + 2^1 + 2^2 + 2^3 + 2^4$.

- My hundreds digit is equal to the sum of the first three odd numbers.

- The sum of all of my digits is the y-intercept of this line: $-5x + y = 13$.

What year am I?

Thousands	Hundreds	Tens	Units
_____	_____	_____	_____

61

On June 18 of this year, Sally Ride became the first American woman in space. She functioned as the mission specialist on a 6-day flight of the space shuttle *Challenger*.

Solve this puzzle to learn the year.

- One of the base angles of a parallelogram is 14° greater than the other base angle. The two-digit number formed by my tens and units digits is the same as the number of degrees in the smaller of the two angles.

- My hundreds digit is equal to the value of d in this equation:
$4(d - 7) = d - 1$.

- The sum of all of my digits is equal to the sum of the first six counting numbers.

What year am I?

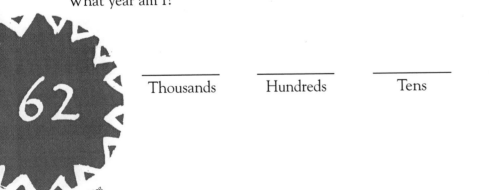

Thousands	Hundreds	Tens	Units

On July 20 of this year, astronaut Neil Armstrong became the first person to land on the moon. When he stepped on the moon, he said, "That's one small step for man, one giant leap for mankind." To learn the year, solve this puzzle.

- The two-digit number formed by my thousands and hundreds digits is the determinant of this matrix:

$$\begin{bmatrix} 5 & 2 \\ -2 & 3 \end{bmatrix}$$

- The two-digit number formed by my tens and units digits is equal to the sum of the whole number greater than or equal to 9 and less than or equal to 14.

- The sum of all of my digits is equal to the value of the function $h(x) = \dfrac{x^2 + 4x - 2}{x - 4}$ for $h(7)$.

What year am I?

63

_____   _____   _____   _____
Thousands   Hundreds   Tens   Units

© 2004 Walch Publishing

On August 17 of this year, three Americans—Max Anderson, Ben Abruzzo, and Larry Newman—became the first to cross the Atlantic in a hot-air balloon. They traveled 3,200 miles in a little over 137 hours. Learn the year by solving this puzzle.

- Objects dropped on Earth fall at a rate of 32 ft/sec^2. The two-digit number formed by my tens and units digits is the distance an object will fall in $2\frac{7}{16}$ sec.

- My hundreds digit is the third square number.

- The sum of all of my digits is $\sqrt{625}$.

What year am I?

64

<u>   </u>   <u>   </u>   <u>   </u>   <u>   </u>

Thousands   Hundreds   Tens   Units

On August 30 of this year, Esther Cleveland became the first baby to be born to the wife of a president in the White House. Learn the year of this famous first by solving this puzzle.

- My tens digit is the positive solution to this problem: $\frac{2}{3}x^2 = 54$.

- Examine the equation $-3x + y = 8$. My units digit is the slope of the line and my hundreds digit is the y-intercept.

- The sum of all of my digits is equal to the median of this set of numbers: 16, 84, 13, 12, 21, 12, 15, 63, 51, 45, 30.

What year am I?

| _____ | _____ | _____ | _____ |
| Thousands | Hundreds | Tens | Units |

65

© 2004 Walch Publishing

On September 25 of this year, the first and only edition of *Public Occurrences Both Foreign and Domestick* was published in Boston. Although this was the first American newspaper, the authorities considered it to be "offensive" and it was suppressed immediately. Solve this puzzle to learn the year.

- The two-digit number formed by my tens and units digits is the percent of people in the United States who in 1990 did not have blood type B.

Blood Type	AB	B	A	O
Number of People	9,950,480	24,876,200	104,480,040	109,455,280
			TOTAL	248,762,000

- My hundreds digit is the number of ways you can roll a sum of 7 using a pair of standard 6-sided dice.

- The sum of all of my digits is equal to $\sqrt[3]{4,096}$.

What year am I?

_____ _____ _____ _____
  Thousands   Hundreds   Tens   Units

On October 2 of this year, Thurgood Marshall was sworn in as the first African-American associate justice of the U.S. Supreme Court. Learn the year by solving this puzzle.

- Benny bought 13 stamps for a total of $3.97. Some cost 37¢ and some cost 25¢. My tens digit is the same as the number of 37¢ stamps; my units digit is the same as the number of 25¢ stamps.

- My hundreds digit is three more than my tens digit; their sum is 15.

- The sum of all of my digits is equal to $3 + 2^3 + 3\sqrt{16}$.

What year am I?

| _____ | _____ | _____ | _____ |
| Thousands | Hundreds | Tens | Units |

On October 25 of this year, the first women to become FBI agents completed their training in Quantico, Virginia. The new agents were Susan Roley and Joanne Pierce. Solve this puzzle to learn the year.

- Walking by a farm you notice that there are chickens and cows in the yard. You count 16 heads and 46 legs. My hundreds digit is the same as the number of chickens; my tens digit is the same as the number of cows.

- My units digit is the slope of this line: $-10x + 5y = 15$.

- The sum of all of my digits is equal to the missing number in this series of hexagonal numbers: 1, 7, __, 37, 61, 91, 127, 169 . . . .

What year am I?

_____   _____   _____   _____
Thousands   Hundreds   Tens   Units

Discoveries, Inventions, and Notable Accomplishments

Looking at a time-line of the history of humankind, one might consider the changes that have come to pass during the twentieth century as nothing less than amazing! My grandmother was born before there were automobiles, my mother was born before there were airplanes, and I was born before there were jet airplanes or men on the moon! In three generations, we have gone from the horse and buggy era to an International Space Station that orbits Earth! In this section, we will look at these discoveries, inventions, and other notable accomplishments that have changed our world. When students solve each of these puzzles, they learn the year when each of these accomplishments occurred and, maybe, they will realize that learning math is really interesting.

On January 5 of this year, German physicist Wilhelm Roentgen announced the discovery of the X-ray. Solve this puzzle to learn the year.

- The length of a rectangle is one less than twice its width. Its area is 45 square units. My units digit is the same as its width; my tens digit its the same as its length.

- My hundreds digit is the area of the entire square if the area of the shaded region is equal to 3.

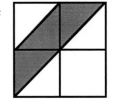

- One hundred twenty-five percent of the sum of all of my digits is equal to 28.75.

What year am I?

_____   _____   _____   _____
Thousands   Hundreds   Tens   Units

© 2004 Walch Publishing

On January 5 of this year, Lizzie Magie received a patent for her board game called "The Landlord's Game." It is very similar to "Monopoly"* except all the properties in Magie's game were rented not purchased. To learn the year Lizzie received her patent, solve this puzzle.

- The area of a parabolic region is equal to $\frac{2}{3}\,bh$. The two-digit number formed by my hundreds and tens digits is equal to the shaded area of this parabola.

- My units digit is equal to the x-coordinate of the vertex of the graph of $y = (x - 4)^2$.

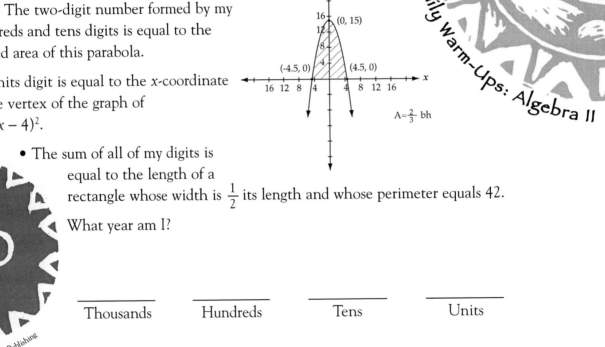

(0, 15)

(-4.5, 0)   (4.5, 0)

$A = \frac{2}{3}\,bh$

- The sum of all of my digits is equal to the length of a rectangle whose width is $\frac{1}{2}$ its length and whose perimeter equals 42.

What year am I?

70

_____ _____ _____ _____
Thousands   Hundreds   Tens   Units

On January 13 of this year, the accordion was patented. This is a musical instrument with keys, metal reeds, and bellows. By fingering the keys, air is forced through the reeds by closing and opening the bellows. Solve this puzzle to learn the year.

- My tens digit is the value of x: $\frac{3}{5} x^2 = 15$.

- Marco bought 12 pencils and pens for \$3.48. Each pencil cost 6¢, and each pen cost 75¢. My hundreds digit is the same as the number of pencils Marco bought, and my units digit is the same as the number of pens.

- The sum of all of my digits is equal to the missing number in this sequence: 1, 3, 4, 7, 11, __, 29 . . . .

What year am I?

_____   _____   _____   _____
Thousands   Hundreds   Tens   Units

71

On January 14 of this year, the Pentagon building in Washington, D.C., was completed. It is the five-sided building that houses the Department of Defense. To learn the year, solve this puzzle.

- Solve this system of equations:

$$\begin{cases} x + y = z - 2 \\ x + y + z = 16 \\ y + x + 1 \end{cases}$$

  My units digit is the same as the value of x; y is the same as my tens digit; z is the same as my hundreds digit.

- The sum of all of my digits is equal to $\sqrt{325 - 36}$.

What year am I?

72

_____   _____   _____   _____
Thousands   Hundreds   Tens   Units

On January 24 of this year, a gold nugget was found at the site of a sawmill owned by John Sutter near Colona, California. This discovery started the California Gold Rush. Learn the year by solving this puzzle.

- Jefferson High School had a bake sale and is presenting a play to raise money for a trip to Washington, D.C. One thousand two hundred sixteen dollars has already been raised at the bake sale. Tickets for the play will be sold for $8. The three-digit number formed by my hundreds, tens, and units digit is the same as the number of tickets that must be sold to raise a total of $8,000.

- The sum of all of my digits is equal to $f(x) = \dfrac{3x}{2} + 9$ for $f(8)$.

What year am I?

_____   _____   _____   _____
Thousands   Hundreds   Tens   Units

73

On February 12 of this year, Alexander Graham Bell demonstrated how his invention, the telephone, worked with a line that ran between Boston and Salem, Massachusetts. Learn the year by solving this puzzle.

- Solve this system of equations:

$$\begin{cases} 2x + 3y = 37 \\ 2x - y = 9 \end{cases}$$

  The x-coordinate of the solution is the same as my hundreds digit; the y-coordinate the same as my tens digit.

- My units digit is the x-intercept of this equation: $3x + 7y = 21$.

- The sum of all of my digits is equal to $(\sqrt{25})^2 - \sqrt{4}$.

  What year am I?

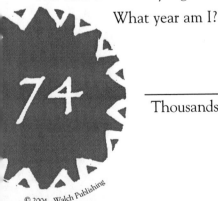

$\underline{\hspace{2cm}}$ $\underline{\hspace{2cm}}$ $\underline{\hspace{2cm}}$ $\underline{\hspace{2cm}}$

Thousands   Hundreds   Tens   Units

Daily Warm-Ups: Algebra II

On February 16 of this year, the tomb of King Tutankhamen was opened by archaeologists. It had been sealed for more than 3,000 years. Learn the year of this discovery by solving this puzzle.

- The two-digit number formed by my thousands and hundreds digits is 4 less than the two-digit number formed by my tens and units digits. Their product is 437.

- The sum of all of my digits is equal to the sum of the first five counting numbers.

What year am I?

_____   _____   _____   _____

Thousands   Hundreds   Tens   Units

75

On February 19 of this year, Thomas Edison received a patent for his invention, the phonograph. It was called the "first talking machine." Recordings at that time were made on wax cylinders. Solve this puzzle to learn the year.

- My hundreds, tens, and units digits form a palindrome with a sum of 23; the product of my even hundreds digit and my odd tens digit is 56.

- The sum of my digits is equal to the sum of the number of degrees in the smallest angle of this triangle:

$4.5x$

x $2x$

What year am I?

Thousands	Hundreds	Tens	Units

76

On March 18 of this year, Schick, Inc. marketed the first electric razor. Learn the year by solving this puzzle.

• The two-digit number formed by my tens and units digit is equal to $7x^6y^{-2} + 3$ when $x = 2$ and $y = 4$.

• My hundreds digit is the GCF of 99 and 117.

• The sum of all of my digits is equal to the value of p in this equation: $\frac{2}{7}\,p + 8 = \frac{1}{2}\,p + 5$.

What year am I?

| _____ | _____ | _____ | _____ |
| Thousands | Hundreds | Tens | Units |

77

©2004 Walch Publishing

On March 17 of this year, paper money became legal tender by an act of Congress. The denominations were $5, $10, and $20. To learn when this money was "invented," solve this puzzle.

- Consider the equation $y = -(x - 2)^2 + 6$. When you graph the parabola, my units digit is the same as the x-coordinate of the vertex; my tens digit is the same as the y-coordinate of the vertex.

- My units digit is $\frac{1}{4}$ of my hundreds digit and $\frac{1}{3}$ of my tens digit.

- The sum of all of my digits is the number you start with in this problem: "Take the sum of the digits, subtract 15, then multiply the difference by 8; take the positive square root of the product; subtract 4. Your answer will be 0.

What year am I?

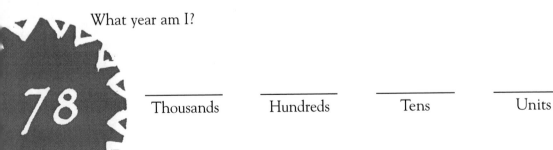

78

_____   _____   _____   _____
Thousands   Hundreds   Tens   Units

On March 26 of this year, Dr. Jonas Salk introduced the polio vaccine in the United States. To learn the year this medical discovery helped wipe out this terrible disease, solve this puzzle.

- Joanna received the following grades on her first four math tests: 96, 85, 92, and 87. The two-digit number formed by my hundreds and tens digits is the grade she will need to get on the fifth test in order to have a 91 average.

- My units digit is the axis of symmetry for the quadratic equation $y = (x - 3)^2$.

- A horse has won 1 race of its last 19 races. The odds against the horse winning the race today are $x : 1$. The sum of all of my digits is equal to the value of x.

What year am I?

_____ _____ _____ _____
Thousands   Hundreds   Tens   Units

79

On April 6 of this year, Teflon® was invented by Dr. Roy Plunkett at the DuPont research laboratories. The surface of Teflon® is so slippery that virtually nothing sticks to or is absorbed by it. To learn the year of this discovery, solve this puzzle.

- A 70-year-old man has a 46-year-old daughter. The two-digit number formed by my tens and units digit is the number of years ago he was four times his daughter's age.

- My hundreds digit is 300% of my tens digit.

- The sum of all of my digits is equal to the sum of the first six counting numbers.

What year am I?

_____	_____	_____	_____
Thousands	Hundreds	Tens	Units

On April 20 of this year, Marie and Pierre Curie isolated the element radium. This husband-and-wife team was awarded the Nobel Prize for their work. Later, Marie Curie won a second Nobel Prize for her work in chemistry, making her the first person to win this prize twice. Even more amazing, Marie Curie's daughter, Irène Joliot-Curie, was also awarded a Nobel Prize. Solve this puzzle to learn the year radium was discovered.

• The two-digit number formed by my thousands and hundreds digits is the largest of the three prime factors of 266.

• My units digit is the smallest of the three prime factors of 266.

• The sum of all of my digits is equal to $\sqrt{2^4 \cdot 3^2}$.

What year am I?

_____   _____   _____   _____
Thousands   Hundreds   Tens   Units

81

© 2004 Walch Publishing

In this year, Ruth Handler, the cofounder of Mattel toys, introduced her invention, the Barbie® doll. She named the doll after her daughter, Barbara. The first doll sold for only $3.00. To learn the year this doll was invented, just solve this puzzle.

- The Acme Taxi Company's rates are $1.25 for the first minute and 30¢ for each additional minute. The two-digit number formed by my tens and units digits is the number of minutes a trip would take for the fare to be $18.65.

- My hundreds digit is the y-intercept of this equation: $3x + 2y = 18$.

- The sum of all of my digits is equal to the number of different ways that four books can be arranged on a shelf.

What year am I?

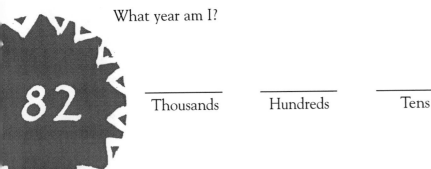

82

_____ Thousands   _____ Hundreds   _____ Tens   _____ Units

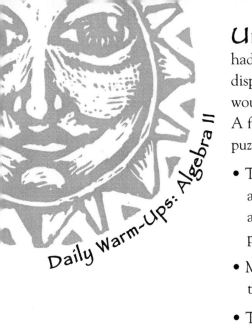

Unhappy with cloth diapers that leaked and had to be washed, Marion Donovan invented the convenient disposable diaper in this year. When companies thought her product would be too expensive to produce, she went into business for herself. A few years later, she sold her business for $1 million. Solve this puzzle to learn the year the disposable diaper was invented.

- The two-digit number formed by my tens and units digits is the same as the shaded area indicated. (Remember, the area of a parabolic section is equal to $\frac{2}{3} bh$.)

- My hundreds digit is one less than twice my tens digit.

- The sum of all of my digits is equal to the sum of the first five counting numbers.

What year am I?

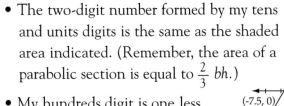

$A = \frac{2}{3}bh$

(0, 5)

(-7.5, 0)

(7.5, 0)

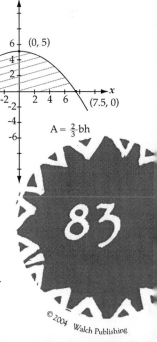

83

Thousands   Hundreds   Tens   Units

Noah McVicker and Joseph McVicker

invented Play-Doh in this year. It received U.S. Patent No. 3,167,440 nine years later. Over 700 million pounds of Play-Doh have been sold. To learn the year this toy was invented, solve this puzzle.

- Solve this system of equations: My hundreds digit is the same as the value of x; my tens digit is the same as the value of y; my units digit is the same as the value of z.

$$\begin{cases} x + y + z = 20 \\ y + z = x + 2 \\ z = y + 1 \end{cases}$$

- The sum of all of my digits is equal to the longer side of a rectangle, the length of which is 6 less than 3 times the width and the area is 189 units^2.

What year am I?

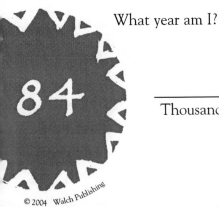

Thousands	Hundreds	Tens	Units

On April 26 of this year, Sarah Boone, one of the first African-American women to receive a patent, invented an improved ironing board. To learn the year of this invention, solve this puzzle.

- The denominator of a fraction is 7 more than the numerator. If both the numerator and denominator were increased by 12, the resulting fraction would equal $\frac{2}{3}$. The numerator of the fraction is equal to my units digit; the denominator is equal to my tens digit.

- My hundreds digit is the solution to the function $f(x) = \dfrac{x^5}{x^2}$ for $f(2)$.

- The sum of all of my digits is the sum of the even numbers greater than or equal to 2 and less than or equal to 8.

What year am I?

————————— ————————— ————————— —————————
Thousands   Hundreds   Tens   Units

85

Richard and Betty James invented the Slinky® in this year. The Slinky® is made from 80 feet of wire, and over a quarter of a billion Slinkys® have been sold worldwide. To learn the year this toy was invented, solve this puzzle.

- My hundreds digit is equal to the number of liters of a 15% solution that must be mixed with 18 liters of a 6% solution to produce a new solution that is 9% acid.

- The two-digit number formed by my tens and units digits is equal to the determinant of this matrix:

$$\begin{bmatrix} 9 & 3 \\ 3 & 6 \end{bmatrix}$$

- You have saved $2.40 in nickels and dimes. Altogether you have 29 coins. The sum of all of my digits is equal to the number of dimes in your collection.

What year am I?

86

_____   _____   _____   _____
Thousands   Hundreds   Tens   Units

On May 19 of this year, the Simplon Tunnel connecting Switzerland and Italy was officially opened. To learn the year, solve this puzzle.

- Solve this system of equations:

$$\begin{cases} y = 1.5x \\ 2x + y = 21 \end{cases}$$

My hundreds digit is equal to the value of y; my units digit is equal to the value of x.

- The two-digit number formed by my hundreds and tens digit is equal to the sum of the even numbers ≥ 2 and ≤ 18.

- The sum of all of my digits is equal to $2^{(4+2)} \cdot 2^{-2}$.

What year am I?

_____ _____ _____ _____
Thousands   Hundreds   Tens   Units

On May 20 of this year, 25-year-old aviator Captain Charles Lindbergh took off from a rainy field in New York to begin the first solitary flight across the Atlantic. His monoplane was called the *Spirit of St. Louis*, and the 3,600-mile flight took almost 39 hours. Solve this puzzle to learn the year.

- The two-digit number formed by my tens and units digits is 3 times my hundreds digit. Their sum is 36.

- Ken has a brine solution in which 18 pounds of salt has been mixed with 114 gallons of water. The sum of all of my digits is equal to the amount of water Ken would mix with 3 pounds of salt to maintain the same proportions.

What year am I?

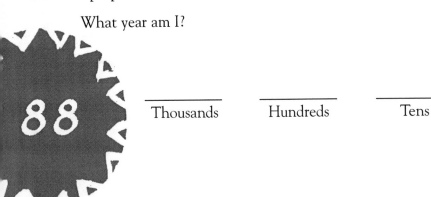

| _____ | _____ | _____ | _____ |
| Thousands | Hundreds | Tens | Units |

On May 24 of this year, the Brooklyn Bridge opened in New York City. It took 14 years to construct and cost over $16 million; it crosses the East River, connecting Brooklyn and Manhattan. This steel suspension bridge was designed by John Roebling and has a span of 1,595 feet. Solve this puzzle to learn the year.

- Each of the two congruent sides of an isosceles triangle are 2 greater than twice the third side. One less than 3 times the third side is equal to the length of a congruent side. My hundreds and tens digits are the same as the length of one of the congruent sides; my units digit is the same as the length of the third side.

- The sum of all of my digits is equal to the number of faces in an icosahedron.

What year am I?

89

_____   _____   _____   _____
Thousands   Hundreds   Tens   Units

Bette Nesmith Graham, a secretary in Dallas, Texas, invented liquid paper in this year. She mixed up the first batch in her home blender to correct her typing errors. She called it "Mistake Out" and started the business out of the basement in her home. Thirty-four years later, she sold the business for $47.5 million. Solve this puzzle to learn the year Bette invented "Mistake Out."

- My tens digit is 1 less than than my units digit. If the digits are reversed, the resulting number is 1 less than 6 times the sum of the digits.

- My hundreds digit is equal to the sum of the first three odd numbers.

- The sum of all of my digits is equal to the total number of spots on one die.

What year am I?

90

| Thousands | Hundreds | Tens | Units |

Daily Warm-Ups: Algebra II

On May 28 of this year, the Dionne quintuplets (Marie, Cecile, Yvonne, Emilie, and Annette) were born in Ontario, Canada. They were the first quints known to survive for more than a few hours after birth. Learn the year the quints were born by solving this puzzle.

- The degree measure of one of two supplementary angles is 10 more than 4 times the other. The two-digit number formed by my tens and units digit is equal to the degree measure of the smaller angle.

- The two-digit number formed by my thousands and hundreds digit is the determinant of this matrix:

$$\begin{bmatrix} 5 & -2 \\ 2 & 3 \end{bmatrix}$$

- The sum of all of my digits is $\log_{10}100{,}000{,}000{,}000{,}000{,}000$.

What year am I?

91

_____ _____ _____ _____
Thousands   Hundreds   Tens   Units

On June 3 of this year, the famous poem "Casey at the Bat" was first printed in the *San Francisco Examiner*. It was written by Ernest Thayer. Solve this puzzle to learn the year.

- My hundreds, tens, and units digits are all the same digit. They are equal to the value of x in this equation: $\log_2 x = 3$.

- The sum of all of my digits is equal to $\sqrt{9^2} + \sqrt{16^2}$.

What year am I?

_____   _____   _____   _____
Thousands   Hundreds   Tens   Units

92

On June 8 of this year, John McGaffrey patented the first suction-type vacuum cleaner. To learn the year, solve this puzzle.

- In the two-digit number formed by my tens and units digits, the tens digit minus my units digit is equal to −3. If the digits are reversed, 6 times the sum of the digits is equal to the new number minus 6.

- My hundreds digit is equal to the value of x: $\log_2 256 = x$.

- The sum of all of my digits is equal to the value of y: $\dfrac{1.2 \times 10^2}{y} = 5$.

What year am I?

Thousands	Hundreds	Tens	Units
_____	_____	_____	_____

On June 15 of this year, Benjamin Franklin conducted his experiment with a kite and proved that lightning contained electricity. To learn the year of this *electrifying* discovery, solve this puzzle.

- In the equation $y = -2(x - 5)^2 + 2$, my tens digit is the x-coordinate of the vertex, and my units digit is the y-coordinate of the vertex.

- My hundreds digit is the shorter side of a rectangle the length of which is 2 greater than the width and the area of which is equal to 63 square units.

- The sum of all of my digits is the discriminant, $(b^2 - 4ac)$, of the quadratic equation $y = \frac{1}{2} x^2 + 4x + 1$.

What year am I?

| Thousands | Hundreds | Tens | Units |

Daily warm-Ups: Algebra II

On June 21 of this year, the wheat reaper was patented. Before this invention, only about 3 acres of wheat could be harvested each day; with the reaper, about 15 could be harvested. (What percent of increase does this represent?) Learn the year of this patent by solving this problem.

- My units digit is equal to the value of m: $\log_m 64 = 3$.

- The sum of the perimeters of a square and an equilateral triangle is 36. One side of the equilateral triangle is two more than twice one side of the square. My tens digit is equal to one side of the square; my hundreds digit is equal to one side of the equilateral triangle.

- The sum of all of my digits is equal to $x^7 \times x^{-3}$ when $x = 2$.

What year am I?

_____   _____   _____   _____
Thousands   Hundreds   Tens   Units

95

Scientist Patsy Sherman had a very fortunate

accident in this year. While trying to develop a rubber hose to use in jet aircraft, she discovered a substance that repelled both water and oil. This substance is now known as Scotchgard™ Fabric Protector. Learn the year this fortunate accident occurred by solving this puzzle.

- My units digit is equal to the value of x: $\log_9 81 = x$.

- The two-digit number formed by my thousands and hundreds digit is the value of this expression when $c = 20$ and $d = 8$: $\frac{1}{2}c + \frac{5}{2}d - 11$.

- My tens digit is equal to the value of x if the perimeter of the triangle equals 23:

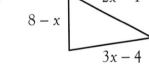

What year am I?

| Thousands | Hundreds | Tens | Units |

Joseph Woodland and Bernard Silver

received a patent for their invention, the bar code, in this year. A bar code is used for automatic identification and data collection—it eventually became what we now call a U.P.C. symbol (Universal Product Code). Twenty-four years after it was invented, the first U.P.C. scanner was used in a store. Solve this puzzle to learn the year of this invention.

- The line that passes through the points (2, 12) and (−4, −18) has a slope that is equal to my tens digit and a y-intercept that is equal to my units digit.

- My hundreds digit is the value of x:

$2x + 3$

x | Area = 135

- The sum of my digits is the value of x:
  $\frac{1}{3}(x - 5) + \frac{3}{2}(x + 1) = 31$.

What year am I?

97

_____   _____   _____   _____
Thousands   Hundreds   Tens   Units

On November 13 of this year, the Holland Tunnel was completed in New York. Running under the Hudson River, it connects Manhattan with Jersey City, New Jersey. Solve this puzzle to learn the year.

- There is a 3° drop in temperature for every 1,000 feet increase in altitude. The two-digit number formed by my tens and units digits is the number of degree drop in temperature at an altitude increase of 9,000 feet.

- My hundreds digit is equal to the value of r in this equation:
  $3r - (-5) = 8[r + (-5)]$.

- A thermometer reads 91.8° Fahrenheit. The sum of all of my digits is equal to the equivalent temperature measured on the Celsius scale. (*Hint:* $C = \frac{5}{9}F - 32°$)

  What year am I?

_____ _____ _____ _____
Thousands   Hundreds   Tens   Units

On November 2 of this year, the largest airplane ever made took its one and only flight over Long Beach Harbor in California. The 200-ton plywood craft, which cost $25 million to build, was named the *Spruce Goose*. Learn the year of this flight by solving this puzzle.

• Solve this system of equations:

$$\begin{cases} 2x + y = 2z + 8 \\ x + y + z = 20 \\ y = z - 3 \end{cases}$$

My hundreds digit is equal to the value of x; my tens digit is equal to the value of y; my units digit is equal to the value of z.

• The sum of all of my digits is the missing number in this sequence: 1, 1, 2, 3, 5, 8, 13,_____, 34, 55. . . .

What year am I?

99

_____   _____   _____   _____
  Thousands   Hundreds   Tens   Units

On November 25 of this year, the patent application for evaporated milk was submitted. Evaporated milk is unsweetened milk that has been thickened by partial evaporation; it is packaged in cans and does not need refrigeration until it is opened. Solve this puzzle to learn the year this patent was submitted.

- My hundreds and tens digit are the same number; they are equal to the value of m: $\log_m 64 = 2$.

- My units digit raised to the third power is equal to the product of my hundreds and tens digits.

- The sum of all of my digits is equal to $\sqrt{500 - 59}$.

    What year am I?

Thousands	Hundreds	Tens	Units

On December 4 of this year, a painting in the Metropolitan Museum of Art in New York was found to have been hung upside down. It had been in this embarrassing position for 47 days. Solve this puzzle to learn the year.

- In this equation, $4x + 6y = 36$; my hundreds digit is the x-intercept; my tens digit is the y-intercept.

- My units digit is the slope of the line containing the coordinates $(0, 4)$ and $(3, 7)$.

- The sum of all of my digits is the y-coordinate of the vertex of the graph of $y = -|x - 4| + 17$.

What year am I?

| _____ | _____ | _____ | _____ |
| Thousands | Hundreds | Tens | Units |

101

© 2004 Walch Publishing

On December 20 of this year, the Louisiana Purchase took place. In one of the greatest real estate deals in history, the United States purchased more than a million square miles of land from France for about $20 per square mile. Learn the year of this purchase by solving this puzzle.

- My units digit is equal to the x-coordinate of the vertex of the graph of $y = |x - 3|$.

- When at rest, we inhale about $12\frac{1}{2}$ liters of oxygen per minute. The three-digit number formed by my thousands, hundreds, and tens digits is the amount of oxygen absorbed in 14 min. 24 sec.

- The sum of all of my digits is equal to the smaller solution of the value of m: $\left|\frac{1}{4}m - 9\right| = 6$.

    What year am I?

Daily Warm-Ups: Algebra II

102

_____   _____   _____   _____
Thousands   Hundreds   Tens   Units

Happy Birthday to You

The puzzles in this section all relate to the birthdays of famous (and sometimes, not so famous) people. Because many of these people will be familiar to students, they may analyze the clues by focusing on the *historical context* rather than on the *isolated* mathematics problem. It is possible you will hear, "That answer doesn't make sense—this person had to be born in the 1900's!" This is a good thing! When students start to look at mathematics as a "sense-making experience," they begin to understand the power of mathematics, and they learn to be real problem-solvers.

On January 1 of this year, American patriot Paul Revere was born in Boston, Massachusetts. He is best known through Longfellow's poem "The Midnight Ride of Paul Revere." Solve this puzzle to learn the year.

- A small plane is 140 miles from its destination at 1:40 P.M. At 2:10 P.M. it is only 52.5 miles from its destination. The three-digit number formed by my thousands, tens, and units digits is the average speed of the plane.

- If $x = 0$, then $x - (-3)$ is equal to my tens digit.

- The sum of all of my digits is equal to the value of x: $\log_2 x = 4$.

What year am I?

_____   _____   _____   _____
Thousands   Hundreds   Tens   Units

103

© 2004 Walch Publishing

On January 3 of this year, author J.R.R. Tolkien was born in South Africa. His best-known works are *The Hobbit* and *The Lord of the Rings* trilogy. To learn the year this creative man was born, solve this puzzle.

- Solve this system of equations.

$$\begin{cases} -2x + 4y = 20 \\ 3x - 3y = -3 \end{cases}$$

My hundreds digit is equal to the value of x; my tens digit is equal to the value of y.

- My units digit is equal to the value of n: $\log_5 25 = n$.
  - The sum of all of my digits is equal to $\begin{pmatrix} 6 \\ 3 \end{pmatrix}$.

  What year am I?

104

_____   _____   _____   _____
    Thousands   Hundreds   Tens   Units

On January 15 of this year, Dr. Martin Luther King, Jr., was born. Dr. King was America's most famous civil rights leader and received the Nobel Peace Prize for his important work. To learn the year of his birth, just solve this problem.

- You have 81 feet of fencing. My hundreds and units digits are the lengths of a rectangular region with the greatest area that you can enclose with this length of fencing (if the sides are whole numbers of feet).

- My tens digit is 126,720 inches rewritten as miles.

- The sum of my digits is $\begin{pmatrix} 7 \\ 2 \end{pmatrix}$.

What year am I?

| _____ | _____ | _____ | _____ |
| Thousands | Hundreds | Tens | Units |

105

© 2004 Walch Publishing

On January 18 of this year, A.A. Milne was born in London, England. He is best known for the children's books *Winnie the Pooh* and *The House at Pooh Corner*. To learn the year he was born, solve this puzzle.

- The two-digit number formed by my tens and units digits is the determinant of this matrix:

$$\begin{bmatrix} 8 & 5 \\ -2 & 9 \end{bmatrix}$$

- Twenty-seven is 150% of the two-digit number formed by my thousands and hundreds digits.

- The sum of all of my digits is equal to $-\frac{1}{3}x^2 + 6x - 5$ when $x = 12$.

What year am I?

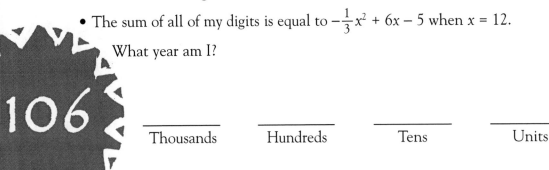

106

_____   _____   _____   _____
Thousands   Hundreds   Tens   Units

On January 24 of this year, Native-American ballerina Maria Tallchief was born. To learn the year of her birth, solve this puzzle.

- The two-digit number formed by my tens and units digit is equal to the value of m: $\log_5 m = 2$.

- My hundreds digit is equal to the slope of this line: $-18x + 2y = 6$.

- The sum of all of my digits is equal to the area of this triangle:

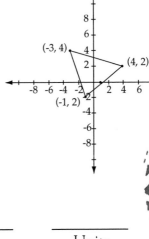

(-3, 4) (4, 2) (-1, 2)

What year am I?

—————————   —————————   —————————   —————————
Thousands   Hundreds   Tens   Units

On January 27 of this year, composer Wolfgang Amadeus Mozart was born in Salzburg, Austria. He began performing at the age of 3 and composed his first piece at the age of 5. He died at the age of 35. To learn the year he was born, solve this puzzle.

- Maria, Janie, and Charlene are friends. The product of Maria's and Janie's ages is equal to 5 less than 5 times Charlene's age. The sum of their ages is 18. Charlene's age minus Maria's age is equal to Janie's age minus 4. Charlene's age is my hundreds digit, Maria's my tens digit, and Janie's my units digit.

- The sum of all of my digits is equal to n: $\log_n 361 = 2$.

What year am I?

Thousands	Hundreds	Tens	Units
_____	_____	____	____

On February 5 of this year, African-American baseball player Hank Aaron was born in Mobile, Alabama. Aaron topped Babe Ruth's batting average and had a career total of 755 home runs. Solve this puzzle to learn the year this baseball great was born.

- The two-digit number formed by my tens and units digits is coterminal with a 394° angle.

- My hundreds digit is equal to $\frac{1^{-2}}{3}$.

- The sum of all of my digits is the median of this set of data: 27, 18, 3, 36, 9, 16.

What year am I?

_____	_____	_____	_____
Thousands	Hundreds	Tens	Units

109

On February 7 of this year, Laura Ingalls Wilder was born. The author of the "Little House" series, she did not begin writing books until she was 65 years old.

- My units digit is the positive solution of $5x^2 = 245$.

- My tens digit is equal to $\begin{pmatrix} 4 \\ 2 \end{pmatrix}$.

- The two-digit number formed by my thousands and hundreds digit is the determinant of this matrix:

$$\begin{bmatrix} 3 & -3 \\ 2 & 4 \end{bmatrix}$$

What year am I?

110

_____   _____   _____   _____
Thousands   Hundreds   Tens   Units

On February 13 of this year, American artist Grant Wood was born in Anamosa, Iowa. His most famous painting is called *American Gothic*. (Can you describe this painting?)

- My hundreds digit is the width of this rectangle; my tens digit is the length:

$$x \quad \boxed{\text{Area} = 72 \ u^2}$$
$$x + 1$$

- My units digit is the value of p: $\log_7 49 = p$.

- Sixteen is $\frac{4}{5}$ of the sum of my digits.

What year am I?

Thousands	Hundreds	Tens	Units
_____	_____	_____	_____

111

On February 15 of this year, astronomer Galileo was born in Pisa, Italy. Galileo is credited with introducing the "scientific method" and was ridiculed for exploring unpopular scientific theories. He said, "We cannot discover new oceans unless we have the courage to lose sight of the shore."

- The two-digit number formed by my thousands and hundreds digits is equal to the longer side of this rectangle, the area of which is 150 square units.

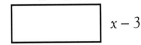

$x + 2$

$x - 3$

- The two-digit number formed by my tens and units digits is equal the value of r in this equation: $\frac{3}{4}r = \frac{5}{8}r + 8$.

- The sum of all of my digits is equal to n: $\log_2 n = 4$.

What year am I?

112

_____ _____ _____ _____
Thousands   Hundreds   Tens   Units

Daily Warm-Ups: Algebra II

On February 15 of this year, Susan B. Anthony was born.

The leader of the women's suffrage movement, she was arrested and fined for voting—which at the time was illegal for women. She was the first woman to appear on an American coin. To learn the year she was born, solve this puzzle.

- My hundreds digit is equal to the diameter of a circle with the equation $x^2 + y^2 = 4$.

- The two-digit number formed by my tens and units digits is the complement of a 70° angle.

- The sum of all of my digits is the value of w: $\frac{1}{5}(4w - 4) = w - 3$.

What year am I?

——————————   ——————————   ——————————   ——————————

Thousands   Hundreds   Tens   Units

113

© 2004 Walch Publishing

On February 18 of this year, Russian-born author and humorist Shalom Aleichem (the pen name for Solomon Rabinowitz) was born in the Ukraine. He was affectionately known in the United States as the "Jewish Mark Twain." Solve this puzzle to learn his birth year.

- My units digit is equal to $\sqrt[6]{3^{12}}$.
- My hundreds digit is the value of d in this equation: $d^{\frac{2}{3}} = 4$.
- My tens digit is equal to the value of n: $\log_n 25 = 2$.
- The sum of all of my digits is equal to the value of x: $\frac{1}{3}(x-2) = x - 16$.

What year am I?

| Thousands | Hundreds | Tens | Units |

On February 19 of this year, Polish astronomer Nicolaus Copernicus was born. He revolutionized scientific thought, arguing that the sun was at the center of our planetary system and that the earth revolved around it.

- My hundreds digit is equal to the value of x: $\sqrt{\sqrt{2}} = 2^{\frac{1}{x}}$.

- My units digit is the value of x: $a^{\frac{2}{6}} b^{\frac{2}{6}} = \sqrt[x]{ab}$.

- My tens digit is the determinant of this matrix:

$$\begin{bmatrix} -5 & 2 \\ -11 & 3 \end{bmatrix}$$

- The sum of my digits is equal to $_6C_2$

What year am I?

$\underline{\hspace{3cm}}$	$\underline{\hspace{3cm}}$	$\underline{\hspace{3cm}}$	$\underline{\hspace{3cm}}$
Thousands	Hundreds	Tens	Units

115

© 2004 Walch Publishing

On February 25 of this year, French Impressionist painter Pierre-Auguste Renoir was born in Limoges, France. Renoir is noted for his delicate use of color and light. To learn the year he was born, solve this puzzle.

- My thousands and units digits are the same number. They are the solution to this expression, when $x = 2$ and $y = 3$: $(2x^2y^3)^0$.

- The two-digit number formed by my hundreds and tens digits is equal to $\binom{9}{3}$.

- The sum of all of my digits is equal to an angle that is coterminal with a 374° angle.

What year am I?

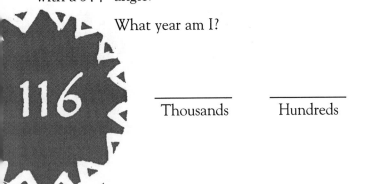

116

_____   _____   _____   _____
Thousands   Hundreds   Tens   Units

On March 2 of this year, author and illustrator Dr. Seuss (Theodore Seuss Geisel) was born in Springfield, Massachusetts. (Can you name some of his popular books?)

- My units digit is equal to the value of n: $\log_3 81 = n$.

- My hundreds digit is the slope of the line that passes through the points $(1, -8)$ and $(3, 10)$.

- The sum of all of my digits is equal to the y-intercept of the line that passes through the points $(-1, 11)$ and $(-3, 5)$.

What year am I?

_____   _____   _____   _____
    Thousands   Hundreds   Tens   Units

117

© 2004 Walch Publishing

On March 6 of this year, painter and sculptor Michelangelo (Buonarroti) was born in the town of Caprese, Tuscany. He is famous for his painting of the Sistene Chapel in Rome and his statue of David (which is located in Florence). Solve this puzzle to learn his birth year.

- The two-digit number formed by my thousands and hundreds digits is equal to the area of the shaded region:

- The two-digit number formed by my tens and units digits is equal to the value of x: $3(x - 8) - 2(x + 5) = 41$.

- The sum of all of my digits is equal to the complement of a 73° angle.

What year am I?

Thousands	Hundreds	Tens	Units

On March 7 of this year, American naturalist Luther Burbank was born in Lancaster, Massachusetts. He is the creator and developer of many new varieties of flowers, fruits, vegetables, and trees. His birthday is celebrated as Bird and Arbor Day. Solve this puzzle to learn the year Burbank was born.

- My hundreds digit is the area of this triangle:

- The two-digit number formed by my tens and units digits is equal to the value of w: $\log_7 w = 2$.

- Harry earned \$200 last week. After 7 days, he had \$46 left. The sum of all of my digits is the average amount Harry spent per day.

What year am I?

_____   _____   _____   _____
Thousands   Hundreds   Tens   Units

119

On March 14 of this year, Albert Einstein was born in Ulm, Germany. A theoretical physicist, he is best known for the theory of relativity. He immigrated to the United States and won the Nobel Prize for his work. Learn the year Einstein was born by solving this puzzle.

- One of the nonright angles of a right triangle is 4 times the other. The two-digit number formed by my thousands and hundreds digits is the smaller of these two angles.

- Solve the system of equations.

$$\begin{cases} \dfrac{1}{3}x - y = -4 \\ 2y - x = 5 \end{cases}$$

My units digit is equal to the value of x; my tens digit is equal to the value of y.

- The sum of my digits is equal to the value of s: $\log_5 s = 2$.

What year am I?

120

_____	_____	_____	_____
Thousands	Hundreds	Tens	Units

On March 21 of this year, Mexican resistance hero Pablo Juárez was born to Zapotec Indian parents. Orphaned at an early age, he became the symbol of Mexican resistance to foreign intervention. Learn the year Juárez was born by solving this puzzle.

- My units digit is equal to the value of n: $\sqrt[3]{\sqrt{2}} = 2^{\frac{1}{n}}$.

- My hundreds digit is the y-coordinate of the vertex of the graph of $y = (x - 3)^2 + 8$.

- The sum of all of my digits is equal to $_6C_2$.

What year am I?

_____	_____	_____	_____
Thousands	Hundreds	Tens	Units

121

On March 23 of this year, mathematician Emmy Noether was born in Erlangen, Germany. She received a doctoral degree in mathematics in Germany. She was forced to leave Germany when the Nazis came into power (she was Jewish); she came to the United States and taught at Bryn Mawr College. Solve this puzzle, and learn the year she was born.

- You have a total of 12 coins; they are nickels and pennies. Altogether you have 28¢. My tens digit is equal to the number of pennies; my units digit is equal to the number of nickels.

- My hundreds digit is equal to the value of x: $\log_2 x = 3$.

- The sum of all of my digits is equal to $\binom{7}{2}$.

    What year am I?

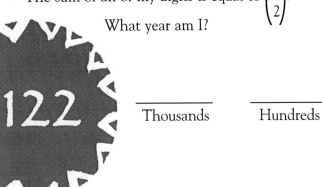

122

Thousands	Hundreds	Tens	Units
___	___	___	___

On March 31 of this year, Franz Joseph Haydn was born in Rohrau, Austria-Hungary. The "Father of the Symphony," he composed 120 symphonies, more than 100 works for chamber groups, a dozen operas, and hundreds of other musical works. Solve this puzzle to learn the year Haydn was born.

- The two-digit number formed by my tens and units digits is equal to the area of the shaded portion:

- My hundreds digit is equal to the determinant of this matrix:

$$\begin{bmatrix} 9 & 5 \\ -5 & -2 \end{bmatrix}$$

- Ten is $83\frac{1}{3}$% of the sum of all of my digits.

What year am I?

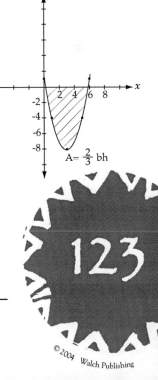

$A = \frac{2}{3}\,bh$

123

Thousands	Hundreds	Tens	Units
___	___	___	___

April 1 is the birthdate of Indian political leader Jagjivan Ram. Born into a family of "untouchables," he was the first of that caste to attend university. A champion and spokesperson for India's 100 million untouchables, he overcame most of the handicaps of the caste system. Learn the year Ram was born by solving this puzzle.

- The two-digit number formed by my hundreds and tens digits is equal to the determinant of the matrix:

$$\begin{bmatrix} 8 & -2 \\ 17 & 7 \end{bmatrix}$$

- My units digit is equal to the value of z: $\frac{3z}{2} = z + 4$.

  - The sum of all of my digits is the percentage that 126 is of 700.

    What year am I?

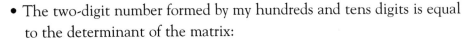

124

_____   _____   _____   _____
    Thousands   Hundreds   Tens   Units

On April 5 of this year, African-American educator and leader Booker T. Washington was born in Franklin County, Virginia. He wrote, "No race can prosper till it learns that there is as much dignity in tilling a field as in writing a poem." Solve this puzzle to learn the year he was born.

- The two-digit number formed by my tens and units digits is equal to $_8C_3$.

- The two-digit number formed by my thousands and hundreds digits is the diameter of the circle with the equation $x^2 + y^2 = 81$.

- The sum of all of my digits is equal to $\binom{6}{3}$.

What year am I?

_____	_____	_____	_____
Thousands	Hundreds	Tens	Units

125

On April 10 of this year, Frances Perkins was born in Boston, Massachusetts. The first woman member of a U.S. presidential cabinet, she was appointed Secretary of Labor by Franklin Delano Roosevelt. Learn the year she was born by solving this puzzle.

- Terry bought 10 notebooks and pens for $7.30. Each notebook cost 79¢ and each pen cost 49¢. My hundreds digit is equal to the number of notebooks and my units digit is equal to the number of pens Terry bought.

- My tens digit is equal to a: $\log_a 512 = 3$.

- The sum of all of my digits is equal to the value of n: $\dfrac{n+2}{3} + \dfrac{n+6}{5} = 12$

What year am I?

126

_____   _____   _____   _____
Thousands   Hundreds   Tens   Units

On April 14 of this year, Anne Sullivan was born. She was the teacher and companion of Helen Keller, the remarkable woman who was both blind and deaf. To learn the year Ms. Sullivan was born, solve this puzzle.

- The length of each side of an equilateral triangle is 2 greater than the length of each side of a square. Both have a perimeter of 24 units. My tens and units digits are the same number and are equal to the length of each side of the square; my hundreds digit is equal to the length of each side of the triangle.

- The sum of all of my digits is equal to 1,3330,560 inches converted to miles.

What year am I?

_____   _____   _____   _____

Thousands   Hundreds   Tens   Units

127

On April 15 of this year, mathematician Leonhard Euler was born. He was the first to use the symbol pi (π), and he published a list of 30 pairs of amicable numbers. (Can you find out what makes a number *amicable*?) Solve this puzzle to find the year Euler was born.

- The two-digit number formed by my hundreds and tens digits is the smallest positive coterminal angle of a 430° angle.

- My units digit is the axis of symmetry for the parabola $y = (x - 7)^2 + 2$.

- Six students are being put into pairs for a project. The number of different pairs that can be formed is equal to the sum of all of my digits.

What year am I?

_____   _____   _____   _____

Thousands   Hundreds   Tens   Units

On April 23 of this year, William Shakespeare was born in Stratford-upon-Avon in England. He wrote at least 36 plays and 154 sonnets. Shakespeare also died on April 23. Solve this puzzle to learn the year he was born.

- My hundreds, tens, and units digits are consecutive integers but are out of order. The sum of the middle and largest is equal to 1 less than 3 times the smallest. My units digit is the smallest, my hundreds digit is the middle, and my tens digit is the largest.

- The sum of all of my digits is equal to the determinant of the matrix:

$$\begin{bmatrix} 0 & -4 \\ 4 & 12 \end{bmatrix}$$

What year am I?

_____   _____   _____   _____

Thousands   Hundreds   Tens   Units

129

On April 26 of this year, Charles Francis Richter was born near Hamilton, Ohio. Richter developed the earthquake magnitude scale that is named after him. If an earthquake has a magnitude of 1 on the Richter scale, it has a power of 10^1, a magnitude of 5 is equal to 10^5 and is 10,000 times more powerful. Find the year he was born.

- The human body absorbs about 0.5 liters of oxygen per minute. The three-digit number formed by my hundreds, tens, and units digits is equal to minutes it will take to absorb 450 liters of oxygen.

- The sum of my digits is 1 million cm converted to km.

What year am I?

130

| _____ | _____ | _____ | _____ |
| Thousands | Hundreds | Tens | Units |

On April 30 of this year, German mathematician Carl Friedrich Gauss was born. When young Gauss was ordered by his teacher to find the sum of all of the numbers from 1 to 100, he found the answer in a flash by recognizing this pattern:

$$1 + 2 + 3 + \ldots\ldots + 98 + 99 + 100$$

$$1 + 100 = 101, \; 2 + 99 = 101, \; 3 + 98 = 101$$

Because there are 50 pairs of 101, the sum is 101 x 50.

- My hundreds, tens, and units digits could be the sides of a cube with a volume of 343 cubic units.

- The sum of all of my digits is equal to the smaller value of m:
$$\left| \frac{1}{2} m - 20 \right| = 9.$$

What year am I?

131

© 2004 Walch Publishing

———————   ———————   ———————   ———————

  Thousands   Hundreds   Tens   Units

On May 12 of this year, English author Edward Lear was born. He is remembered for his limericks, published in a book entitled *Book of Nonsense*. Here's an example:

> There was a young poet named Lear
> Who said it is just as I fear
> Five lines are enough
> For this kind of stuff
> Make a limerick each day of the year.

- The two-digit number formed by my tens and units digits is equal to the area of the triangle with vertices at $(0, 4)$, $(6, 4)$, and $(6, 8)$.

- My hundreds digit is equal to the value of x:

$$\begin{vmatrix} 2 & x \\ 2 & 5 \end{vmatrix} = -6$$

- The sum of all of my integers is $6(8)^{\frac{1}{3}}$.

What year am I?

132

Daily Warm-Ups: Algebra II

_____   _____   _____   _____
Thousands   Hundreds   Tens   Units

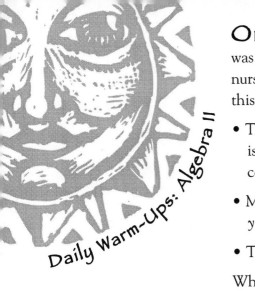

On May 12 of this year, English nurse Florence Nightingale was born. She is considered the person most responsible for making nursing a respected profession. Learn the year she was born by solving this puzzle.

- The two-digit number formed by my thousands and hundreds digits is the number of degrees in the angle when the radian $\frac{\pi}{10}$ is converted to degrees. (*Hint:* $\frac{\pi}{180} = 1°$)

- My tens digit is the y-coordinate of the vertex of the parabola $y = (x + 2)^2 + 2$.

- The sum of all of my digits is 0.2% of 5,500.

What year am I?

| _____ | _____ | _____ | _____ |
| Thousands | Hundreds | Tens | Units |

133

On May 15 of this year, author Lyman Frank Baum was born in Chittenango, New York. An American newpaperman, Baum wrote the Wizard of Oz stories. *The Wonderful World of Oz* was the most famous, but he wrote many other books for children—including more than a dozen about Oz. Find the year he was born.

- The two-digit number formed by my tens and units digits is equal to $7(64\frac{1}{2})$.

- If $f(x) = 2x - 1$ and $g(x) = x + 6$, my hundreds digit is equal to $g(3) - f(1)$.

- The sum of my digits is equal to the positive value of x: $2x^2 = 800$.

What year am I?

134

_____   _____   _____   _____
Thousands   Hundreds   Tens   Units

On May 16 of this year, mathematician Maria Gaetana Agnesi was born in Italy. Maria was a gifted child who spoke more than a dozen languages by the age of 9. Her love, however, was mathematics; and even in a time when women were not educated, she became famous for her published books of mathematics. Learn the year she was born by solving this puzzle.

- My units digit is equal to the x-coordinate of the vertex of the graph of $y = (x - 8)^2 + 1$; my tens digit is the y-coordinate of the vertex of the same graph.

- My hundreds digit is the x-coordinate of the vertex of the graph of $y = -|x - 7| + 2$.

- The sum of my digits is the same as the two-digit number formed by my thousands and hundreds digits.

What year am I?

135

_____   _____   _____   _____
Thousands   Hundreds   Tens   Units

On May 17 of this year, English physician Edward Jenner was born. He is the originator of the inoculation that we use against smallpox. During the Middle Ages, smallpox epidemics caused the death of millions of people. Learn the year Jenner was born by solving this puzzle.

- The two-digit number formed by my tens and units digit is equal to the value of the expression $x^4 \div x^2$ when $x = 7$.

- The two-digit number formed by my thousands and hundreds digit is equal to the value of x: $5(x - 8) = 2x + 11$.

- The sum of all of my digits is equal to $_7C_2$.

What year am I?

Thousands	Hundreds	Tens	Units

On May 19 of this year, African-American civil rights leader Malcolm X (Malcolm Little) was born in Omaha, Nebraska. He took the letter "X" to protest the family name assigned by white slave owners to their slaves. He began the Organization of American Unity. He was assassinated in New York City. Find the year he was born.

- My tens digit is equal to $f(3) + g(3)$ when $f(x) = x - 4$ and $g(x) = x + 6$.

- My hundreds digit is equal to $g(f(1))$.

- My units digit is equal to $f(g(-3))$.

- The sum of all of my digits is equal to $\sqrt{300 - 11}$.

What year am I?

Thousands	Hundreds	Tens	Units

137

On May 23 of this year, German physician Friedrich Anton Mesmer was born. He developed a form of hypnotism, that was called "mesmerism," to treat his patients. Solve this problem to learn the year he was born.

- My units digit is equal to the value of n: $\log_n 1024 = 5$.

- My hundreds digit is equal to the value of x in this system; my tens digit is equal to the value of y.

$$\begin{cases} 2x + y = 17 \\ y = -4 + x \end{cases}$$

- The sum of all of my digits is equal to $3(125)^{\frac{1}{3}}$.

What year am I?

_____ _____ _____ _____
Thousands   Hundreds   Tens   Units

Daily Warm-Ups: Algebra II

© 2004 Walch Publishing

On June 5 of this year, the Greek philosopher Socrates was born. His student Plato wrote down Socrates' ideas and teachings. (We need to write B.C.E. before his birth date.)

- You have the following six items to choose for your lunch: a sandwich, an apple, a fruit snack, yogurt, a cookie, and a juice box. My hundreds digit is equal to the number of combinations there are if you choose to eat only one item; my tens digit is equal to the number of combinations if you choose to eat three of the six items.

- My units digit is 150% of my tens digit.

- The sum of all of my digits is equal to the number of pounds in 304 ounces.

What year am I?

| Thousands | Hundreds | Tens | Units |

139

On June 12 of this year, Anne Frank was born. She was the young Jewish girl who kept a diary while hiding in an attic to avoid being captured by the Nazis. It was later published as a book called *Anne Frank: Diary of a Young Girl.* Solve this puzzle to learn the year Anne was born.

- My hundreds, tens, and units digits form a palindrome with a sum of 20.

- My tens digit is equal to $(\frac{1}{16})^{\frac{-1}{4}}$.

- The sum of all of my digits is equal to the product of the second and fourth prime numbers.

What year am I?

140

_____   _____   _____   _____
Thousands   Hundreds   Tens   Units

On July 22 of this year, American poet Emma Lazarus was born. Words from her poem "The New Colossus" appear at the base of the Statue of Liberty. Some of the words are "Give me your tired, your poor, your huddled masses yearning to be free." Find the year she was born by solving this puzzle.

- My hundreds digit is equal to $4^{\frac{3}{2}}$.

- My tens digit is equal to $(\frac{1}{64})^{\frac{-1}{3}}$.

- My units digit is equal to the value of n: $\log_n 81 = 2$.

- The sum of all of my digits is equal to the determinant of this matrix:

$$\begin{bmatrix} 5 & -2 \\ 6 & 2 \end{bmatrix}$$

What year am I?

_____   _____   _____   _____
Thousands   Hundreds   Tens   Units

On September 24 of this year, puppeteer Jim Hensen was born in Greenville, Mississippi. Among his creations were Muppets Kermit the Frog, Big Bird, Bert and Ernie, Miss Piggy, the Cookie Monster, and Oscar the Grouch. Find the year Jim Hensen was born by solving this puzzle.

- The two-digit number formed by my tens and units digits is equal to the angle measure of $\frac{\pi}{5}$. (*Hint:* $\frac{\pi}{180} = 1°$)

- My tens digit is equal to the value of x: $512^{\frac{1}{x}} = 2$.

- The sum of all of my digits is equal to $x^3 + x^2 + 7$ when $x = 2$.

What year am I?

142

<u>   </u>   <u>   </u>   <u>   </u>   <u>   </u>

Thousands   Hundreds   Tens   Units

On October 3 of this year, rock-and-roll singer Chubby Checker was born in Philadelphia, Pennsylvania. His real name was Ernest Evans. His most famous song was "The Twist." Solve this puzzle to learn the year he was born.

- When this fraction $(\frac{8}{27})^{\frac{2}{3}}$ is simplified, my tens digit is equal to its numerator and my hundreds digit to the denominator.

- My units digit is equal to tan 45°.

- If $f(x) = 2x - 7$ and $g(x) = x - 8$, then the sum of all of my digits is equal to $f(g(3))$.

What year am I?

Thousands	Hundreds	Tens	Units

143

On October 19 of this year, mountain climber Annie Peck was born. She climbed the Matterhorn in the Swiss Alps and became the first American to climb the Peruvian peak Huascarán (22,205 feet). At the age of 61, she placed a banner saying "Votes for Women" at the top of Mount Coropuna in Peru. Solve the puzzle to learn her birth year.

- The height of a triangle is 3 cm greater than the length of its base. Its area is 20 cm². My hundreds digit is equal to the height of this triangle; my tens digit is equal to its base.

- My units digit is equal to the determinant of this matrix:

$$\begin{bmatrix} 4 & -10 \\ -2 & 5 \end{bmatrix}$$

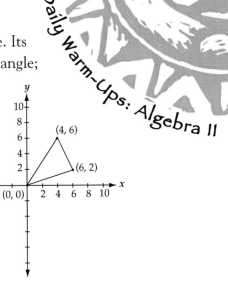

- The sum of all of my digits is equal to the area of this triangle:

  What year am I?

144

| Thousands | Hundreds | Tens | Units |

On October 29 of this year, political cartoonist Bill Mauldin was born. Perhaps his most famous cartoon was printed after the assassination of President Kennedy. It portrayed the Lincoln Memorial in tears. Solve this puzzle to learn the year he was born.

- My tens digit is both the x and y coordinate of the positive point of intersection of the graphs of $x^2 + y^2 = 8$ and $y = \frac{1}{2}x + 1$.

- Solve this system of equations:
$$\begin{cases} x - y + 8 \\ x + 5y = 14 \end{cases}$$

  My hundreds digit is equal to x; my thousands digit is equal to y.

- The sum of all of my digits is equal to s: $\log_s 169 = 2$.

What year am I?

_____   _____   _____   _____
Thousands   Hundreds   Tens   Units

145

On November 8 of this year, English astronomer and mathematician Edmund Halley was born. He was the first to observe the comet that is named after him. Halley's Comet can be seen about every 76 years; it is expected to be visible again in 2061. There have been only 28 recorded appearances since 240 B.C. Find the year Halley was born.

- The three-digit number formed by my hundreds, tens, and units digits is a palindrome that is equal to $64^{1.5} + 12^2$.

- The sum of all of my digits is equal to the area of this triangle:

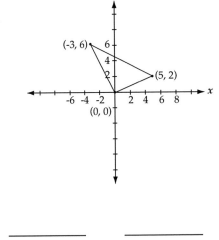

What year am I?

146

Thousands	Hundreds	Tens	Units

On November 17 of this year, German astronomer and mathematician August Möbius was born. He was the pioneer of the field of mathematics known as topology and was the first to describe the Möbius strip. Find the year he was born by solving this puzzle.

- My units digit is equal to cos 90°.

- My hundreds digit is equal to the length of the shorter side of this rectangle; my tens digit to the longer side:

$$\boxed{\text{Area} = 63 \ u^2} \quad x$$
$$x + 2$$

- Two hundred fifty percent of the sum of all of my digits is equal to 42.5.

What year am I?

147

| Thousands | Hundreds | Tens | Units |

Historical Highlights

From the first election for president of the United States to the radio broadcast of *The Lone Ranger,* these puzzles encourage students to explore history from a *mathematical perspective.* Exponents, square roots, number patterns, and more are used to help students find, among other things, the year of the highest wind speed or the last time M*A*S*H was seen on television.

Students will again be using context clues to help them solve these puzzles. Of course, television shows will be shown in the twenty-first century—relating historical events to abstract mathematics clues take the *mystery* out of the problem and will help students through the frightful "story problem syndrome."

On January 7 of this year, the first presidential election was held in the United States. To learn the year of this *historical highlight* solve this puzzle.

- When you solve this system of equations, x is equal to my hundreds digit, y is equal to my tens digits, and z is equal to my units digit:

$$\begin{cases} z - x = 2 \\ x + y + z = 24 \\ y = \frac{1}{2}(x + z) \end{cases}$$

- The sum of all of my digits is equal to the value of x: $\log_5 x = 2$.

What year am I?

_____   _____   _____   _____

Thousands   Hundreds   Tens   Units

Daily Warm-Ups: Algebra II

On January 10 of this year, the League of Nations was established. This was the predecessor of the United Nations and was dissolved 26 years after it was formed. The United States never joined. Learn the year it was established by solving this puzzle.

- My tens digit is the x-coordinate and my hundreds digit is the y-coordinate of the vertex of the graph of $y = (x - 2)^2 + 9$.
- My units digit is equal to the value of this expression: $\frac{18}{3!} - 3$.
- The sum of all of my digits is the area of the triangle with vertices at $(0, 0)$, $(5, -4)$, and $(-1, -4)$.

What year am I?

———————— ———————— ———————— ————————
Thousands   Hundreds   Tens   Units

149

© 2004 Walch Publishing

On January 11 of this year, Chicago schools were closed in the wake of the record-breaking –26°F temperatures. Learn the year by solving this puzzle.

- My units digit is equal to the value of n: $\log_{15} 225 = n$.

- My tens digit is the geometric mean of 16 and 4.

- My hundreds digit is the determinant of the following matrix:

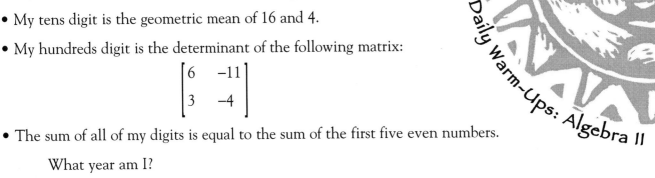

$$\begin{bmatrix} 6 & -11 \\ 3 & -4 \end{bmatrix}$$

- The sum of all of my digits is equal to the sum of the first five even numbers.

  What year am I?

150

_____ Thousands _____ Hundreds _____ Tens _____ Units

On January 19 of this year, the world's record snowfall fell on London, England. Drifts of 4.5 meters were measured. (How many feet is 4.5 meters?) Solve this puzzle to learn the year.

- My hundreds and tens digits could be the radii of the graph of the circle $x^2 + y^2 = 64$.

- My units digit is equal to $\tan 45°$.

- The sum of my digits is the x-coordinate of the solution of this system of equations:

$$\begin{cases} x + y = 27 \\ x - y = 9 \end{cases}$$

What year am I?

_____   _____   _____   _____

Thousands   Hundreds   Tens   Units

151

On January 28 of this year, the world watched in horror as the space shuttle *Challenger* exploded 74 seconds into its flight. The seven people who died were teacher Christa McAuliffe and six other crew members—Francis Scobee, Michael Smith, Judith Resnick, Ellison Onizuka, Ronald McNair, and Gregory Jarvis. Learn the year of this tragedy by solving this puzzle.

- My units digit is the diameter of the graph of $x^2 + y^2 = 9$.

- My tens digit is equal to n: $n^{\frac{2}{3}} = 4$.

- My hundreds digit is equal to x: $\sqrt{x^2} - 2 = 7$.

- The sum of all of my digits is the geometric mean of 64 and 9.

What year am I?

152

_____   _____   _____   _____
Thousands   Hundreds   Tens   Units

On January 30 of this year, the first radio broadcast of *The Lone Ranger* was heard in the United States. The song played at the beginning of the program was the "William Tell Overture" from Rossini's opera. The program started and ended with the phrase "Heigh-ho, Silver." Learn the year by solving this puzzle.

- My tens digit and units digit are the same. They are equal to the value of a: $\dfrac{a + 3}{4} = \dfrac{a}{2}$.

- My hundreds digit is equal to the value of w: $\log_w 729 = 3$.

- If $f(x) = x^2$ and $g(x) = x + 2$, then the sum of all of my digits is equal to $f(g(2))$.

What year am I?

_____ _____ _____ _____
Thousands   Hundreds   Tens   Units

153

On February 1 of this year, the U.S. Supreme Court held its first session in the Royal Exchange Building in New York. Solve this puzzle to learn the year.

- The two-digit number formed by my tens and units digits is equal to the quotient of $\dfrac{2.7 \times 10^5}{3 \times 10^3}$ written in decimal form.

- My hundreds digit is equal to the value of x: $\dfrac{x - 11}{-2} = \dfrac{2x}{7}$.

- The sum of all of my digits is the value of y: $\dfrac{1}{3} = \dfrac{4}{y - 5}$.

What year am I?

154

_____	_____	_____	_____
Thousands	Hundreds	Tens	Units

On February 11 of this year (B.C.E.), the Japanese nation was founded when Emperor Jimmu ascended to the throne. The national holiday is called National Foundation Day, and ceremonies are held with the emperor, empress, and many other dignitaries attending. What is this historical year?

- My hundreds and tens digits are the same. They are equal to the sum of $\dfrac{x + 2}{5} + \dfrac{x - 2}{5}$ when $x = 15$.
- My units digit is the quotient of $\dfrac{x^3 - 27}{x - 3} \div \dfrac{x^2 + 3x + 9}{8x - 24}$ when $x = 3$.
- The sum of my digits is 96 pints converted to gallons.

What year am I?

Thousands	Hundreds	Tens	Units
_____	_____	____	_____

155

On February 23 of this year, members of the 5th U.S.

Marine division planted an American flag atop Mount Suribachi on Iwo Jima. This act was memorialized in a very famous photograph. Learn the year this occurred by solving this puzzle.

- My units digit is equal to $\sqrt[-3]{-625}$.

- My tens digit is equal to $|a|$: $\dfrac{a+7}{3} = \dfrac{a+6}{2}$.

- My hundreds digit is equal to

$$\cfrac{1}{1 - \cfrac{1}{1 - \cfrac{1}{1-9}}}$$

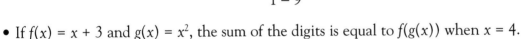

- If $f(x) = x + 3$ and $g(x) = x^2$, the sum of the digits is equal to $f(g(x))$ when $x = 4$.

What year am I?

Thousands	Hundreds	Tens	Units

On February 23 of this year, the siege of the Alamo began in San Antonio, Texas. To learn the year, solve this puzzle.

- My hundreds digit is 2 greater than my units digit. My hundreds digit is equal to the value of s, my units digit to the value of t: t: $\dfrac{s^2}{s-t} - \dfrac{t^2}{s-t} = 14$.
- My tens digit is equal to n: $\log_n 9 = 2$.
- The sum of all of my digits is equal to $\dfrac{0.36 \times 10^5}{0.2 \times 10^4}$.

What year am I?

Thousands	Hundreds	Tens	Units

157

On February 28 of this year, the final episode of the television show M*A*S*H aired. This was the last of the show's 255 episodes and was watched by over 75% of the viewing public. Solve this puzzle to learn the year.

- Solve this system of equations:

$$\begin{cases} 2x + y = 19 \\ y = x - 5 \end{cases}$$

My tens digit is equal to the value of the x-coordinate, my units digit to the value of the y-coordinate.

- My hundreds digit is equal to the axis of symmetry of the graph of the equation $y = (x - 9)^2 + 8$.

- The sum of my digits is equal to $_7C_2$.

What year am I?

_____   _____   _____   _____
  Thousands   Hundreds   Tens   Units

On March 3 of this year, "The Star-Spangled Banner" officially became the national anthem of the United States. To learn the year, solve this puzzle.

- My tens digit is equal to $81^{\frac{1}{4}}$.

- My units digit is equal to tan 45°.

- My hundreds digit is equal to the radius of the graph of the circle $x^2 + y^2 = 81$.

- When $g(x) = 2x^2 - 4$, the sum of all of my digits is equal to $g(3)$.

What year am I?

_____	_____	_____	_____
Thousands	Hundreds	Tens	Units

On March 10 of this year, American abolitionist Harriet Tubman died in Auburn, New York. A former slave, Ms. Tubman escaped from a Maryland plantation and formed the "Underground Railroad." Through her efforts, more than 300 slaves reached freedom. Solve this puzzle to learn the year of her death.

- My units digit is the slope of the line $-3x + y = 2$.

- My tens digit is equal to sin 90°.

- My hundreds digit is 300% of my units digit.

- A person's weight on the moon is directly proportional to their weight on Earth. A 150-lb. man weighs 25 lb. on the moon. The sum of all of my digits is equal to the weight of an 84-lb. boy on the moon.

What year am I?

160

_____   _____   _____   _____
Thousands   Hundreds   Tens   Units

On March 10 of this year, Commissioner George Scott Railaton and seven women officers established the Salvation Army in the United States. To learn the year, solve this puzzle.

- My hundreds and tens digits are the same. They are equal to $4\frac{2}{3}$.

- My units digit is equal to cos 90°.

- The sum of all of my digits is equal to the diameter of the graph of the circle $x^2 + y^2 = 72.25$.

What year am I?

_____	_____	_____	_____
Thousands	Hundreds	Tens	Units

On March 20 of this year, children rang the United Nations bell in New York City for the first celebration of Earth Day. The bell sounded at the exact moment when the sun crossed the equator, the beginning of spring in the Northern Hemisphere. To learn the year, solve this puzzle.

- Find the solution to this system:

$$\begin{cases} x + y = 16 \\ y = x + 2 \end{cases}$$

My tens digit is equal to the value of x; my units digit is equal to the value of y.

- My hundreds digit is the x-coordinate of the vertex of the graph of $y = -(x - 9)^2 + 7$.
- The sum of my digits is equal to the determinant of $\begin{bmatrix} -2 & 8 \\ -4 & 3 \end{bmatrix}$.

  What year am I?

162

_____   _____   _____   _____
Thousands   Hundreds   Tens   Units

On March 23 of this year, Patrick Henry delivered his "Give me liberty or give me death" speech. To learn the year, solve this puzzle.

- My units digit is equal to $625^{\frac{1}{4}}$.

- My hundreds and tens digit are the same. They are equal to n:
  $\log_n 343 = 3$.

- The weight of a person on Earth varies directly as to their weight on Pluto. A 100-lb. person would weigh 6.9 lbs. on Pluto. The sum of all of my digits is the weight of a 2-year-old child on Earth who weighs 1.38 lbs. on Pluto.

What year am I?

Thousands	Hundreds	Tens	Units
_____	_____	_____	_____

163

© 2004 Walch Publishing

On April 11 of this year, the American Society for the Prevention of Cruelty to Animals (A.S.P.C.A.) was chartered in New York State. To learn the year, solve this puzzle.

- My tens and units digits are the same number. They are equal to $7{,}776^{\frac{1}{5}}$.

- The two-digit number formed by my thousands and hundreds digits is the geometric mean of 27 and 12.

- If the length and width of a rectangle were increased by 10%, the percentage of increase in its area would equal the sum of my digits.

What year am I?

164

_____   _____   _____   _____
Thousands   Hundreds   Tens   Units

On April 12 of this year, the wind reached the highest ever recorded in the United States. At the Mount Washington Observatory in New Hampshire, weather observers Wendell Stevenson, Alexander McKenzie, and Salvatore Pagliuca observed and recorded gusts over 231 miles per hour.

- My units digit is equal to the determinant of this matrix:

$$\begin{bmatrix} 0 & 2 & 1 \\ -2 & 0 & 0 \\ 3 & 4 & 3 \end{bmatrix}$$

- My tens digit is the radius of the circle $x^2 + y^2 = 9$.

- My hundreds digit is equal to $27^{\frac{2}{3}}$.

- If $f(x) = x + 2$ and $g(x) = 3x$, then the sum of my digits is equal to $f(g(x))$ when $x = 5$.

What year am I?

_____ | _____ | _____ | _____
Thousands | Hundreds | Tens | Units

165

On April 17 of this year, the 355-year "state of war" that had existed between the Netherlands and the Scilly Isles came to an end. Though hostilities had ended centuries before, no one had bothered to declare the war over. Learn the year of this truce by solving the puzzle.

- Solve this system of equations:

$$\begin{cases} x + y + z = 23 \\ x + y = z + 5 \\ x + z \end{cases}$$

  My units digit is equal to the value of x; my tens digit to the value of y; my hundreds digit to the value of z.

- The sum of all of my digits is equal to 225% of $11\frac{5}{9}$.

  What year am I?

166

Thousands	Hundreds	Tens	Units
_____	_____	_____	_____

On May 6 of this year, the dirigible *Hindenburg* exploded as it approached its mooring in New Jersey. It had just completed a transatlantic crossing. Thirty-six of its ninety-seven passengers and crew died. To learn the year this occurred, solve this puzzle.

- Marcia sold ten tickets to the school play. Adult tickets cost $5 and student tickets cost $3.75. Altogether she sold $46.25 worth of tickets. My units digit is equal to the number of adult tickets, and my tens digit is equal to the number of students tickets.

- My hundreds digit is equal to the value of n: $\log_2 512 = n$.

- The sum of all of my digits is equal to the height of a triangle with a base of 6 and an area of 60 square units.

What year am I?

| _____ | _____ | _____ | _____ |
| Thousands | Hundreds | Tens | Units |

167

On May 7 of this year, a German submarine torpedoed the British passenger ship *Lusitania* while traveling from New York to England. It was carrying nearly 2,000 passengers; 1,198 lives were lost. Germany had warned President Wilson in advance, claiming that the *Lusitania* was carrying ammunition to England. Learn the year by solving this puzzle.

- The two-digit number formed by my tens and units digits is equal to the length of the hypotenuse of the right triangle whose legs are 9 inches and 12 inches.

- My hundreds digit is the numerator of the fraction when this rational number is simplified: $\left(\dfrac{27}{64}\right)^{\frac{2}{3}}$.

  - The sum of all of my digits is equal to the denominator of that same fraction.

  What year am I?

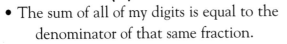

168

Thousands	Hundreds	Tens	Units

On June 2 of this year, Congress granted Native Americans citizenship. Learn the year by solving this puzzle.

- My units digit is 2 greater than my tens digit. The sum of their squares is equal to 20.
- My hundreds digit is equal to the value of r: $\log_r \frac{1}{81} = -2$.
- The sum of all of my digits is the determinant of this matrix:

$$\begin{bmatrix} -2 & -6 \\ 6 & 10 \end{bmatrix}$$

What year am I?

_____ _____ _____ _____
Thousands   Hundreds   Tens   Units

169

On June 15 of this year, King John signed the Magna Carta. This is considered to be one of the most important documents in the history of humanity's search for liberty and freedom. Four copies have survived through the years. Learn the year of this signing by solving this puzzle.

- The two-digit number formed by my tens and units digits is equal to one of the legs of a right triangle whose other leg equals 8 and hypotenuse equals 17.

- My hundreds digit is the positive root of $x^3 - 2x^2 - x + 2 = 0$.

- The sum of all of my digits is equal to $(8)^{\frac{1}{3}} + 7$.

What year am I?

170

| Thousands | Hundreds | Tens | Units |

On July 17 of this year, Douglas Corrigan flew out of Brooklyn, New York, heading for Los Angeles, California. Twenty-eight hours later, he landed in Dublin, Ireland. He was nicknamed "Wrong Way" because he had followed the wrong end of the compass needle. Learn the year of his flight by solving this puzzle.

- To balance a see-saw, the distance (d) from the fulcrum is inversely proportional to a person's weight (w). Marcus weighs 120 lb. and is 6 ft. from the fulcrum. My hundreds digit is the distance Joey must sit from the fulcrum if he weighs only 80 lbs.

- The two-digit number formed by my tens and units digit is the determinant of this matrix:

$$\begin{bmatrix} 2 & 2 \\ -19 & 0 \end{bmatrix}$$

- The sum of all of my digits is equal to $_7C_5$.

What year am I?

171

Thousands	Hundreds	Tens	Units

On August 5 of this year, the first English colony in North America was founded on the Sir Humphrey Gilbert. It was established in the area around St. John's Harbor, Newfoundland. Find the year by solving this puzzle.

- If $f(x) = x + 2$ and $g(x) = 3x$, then my tens digit is equal to $f(g(x))$ when $x = 2$.

- My units digit is equal to $g(f(x))$ when $x = -1$.

- My hundreds digit is equal to $f(g(x))$ when $x = 1$.

- The sum of all of my digits is equal to n: $\log_2(n + 1) = 4$.

What year am I?

172

_____   _____   _____   _____
Thousands   Hundreds   Tens   Units

On August 7 of this year, the first picture of the earth was received from the space satellite *Explorer VI*. It was the first time the people of Earth could see their planet. Learn the year by solving this problem.

- The two-digit number formed by my tens and units digits is equal to x: $\log_4(x - 5) = 3$.

- My hundreds digit is the area of this trapezoid:

- Temperature drops 3° for every 1,000-foot increase in altitude. The sum of my digits is the drop in temperature at an altitude of 8,000 feet.

What year am I?

————— ————— ————— —————
Thousands   Hundreds   Tens   Units

173

On September 23 of this year, the planet Neptune was first observed. Neptune, the eighth planet from the sun, is about 2.8 billion miles from the sun; it takes 165 years to revolve around the sun and has a diameter of about 30,000 miles. (How do these statistics compare with Earth's?) Find the year of this observation.

- The length of a rectangle is 2 inches longer than its width. Its area is 48 sq. in. My hundreds digit is the longer side (length), and my units digit is the shorter side (width).

- My tens digit is equal to x: $\log_x \frac{1}{16} = -2$.

- The sum of my digits is the solution to $3(x - 4) = 2(x + 3\frac{1}{2})$.

    What year am I?

174

_____   _____   _____   _____
  Thousands   Hundreds   Tens   Units

On October 28 of this year, the Statue of Liberty was dedicated on Bedloe's Island in New York. It was sculpted by Frédéric-Auguste Bartholdi who called it *Liberty Enlightening the World*. A sonnet by Emma Lazarus at its base contains the words, "Give me your tired, your poor, your huddled masses yearning to be free." Learn the year of this dedication by solving this puzzle.

- My tens digit is 2 greater than my units digit. The sum of their squares is 100.

- My hundreds digit is equal to n: $\log_2 n = 3$.

- The sum of my digits is the determinant of this matrix:

$$\begin{bmatrix} 2 & -3 \\ 3 & 7 \end{bmatrix}$$

What year am I?

175

_____ _____ _____ _____
Thousands   Hundreds   Tens   Units

On November 9 of this year, New York City and most of the northeastern United States blacked out due to an electric power failure. More than 30 million people, in an 80,000 square mile area, were affected. Solve this puzzle to learn the year that this happened.

- A 100-lb. woman weighs 4 lb. on Pluto. The two-digit number formed by my tens and units digits is the weight of a boy on Earth who weighs 2.6 lb. on Pluto.

- My hundreds digit is the distance between the two points (−4, 2) and (5, 2).

- The sum of my digits is equal to $\begin{pmatrix} 7 \\ 2 \end{pmatrix}$

  What year am I?

_____	_____	_____	_____
Thousands	Hundreds	Tens	Units

On November 18 of this year, the squeaky-voiced Mickey Mouse first appeared on a movie screen at the Colony Theater in New York City. Walt Disney's *Steamboat Willie* was the first animated cartoon talking picture. Learn the year of this movie breakthrough by solving this puzzle.

- A car traveled 133 miles on 7 gallons of gas. The two-digit number formed by my tens and units digits is the number of gallons of gas used by the same car to travel 532 miles.

- My hundreds digit is equal to x: $\sqrt{2x+7} = 5$.

- The sum of my digits is equal to ${}_5P_2$.

What year am I?

<u>   </u>   <u>   </u>   <u>   </u>   <u>   </u>

Thousands   Hundreds   Tens   Units

177

On November 19 of this year, President Abraham Lincoln delivered the Gettysburg Address at a ceremony dedicating 17 acres of the battlefield in Gettysburg, Pennsylvania. Though the speech took less than 2 minutes, it is recognized as one of the most eloquent in the English language. Find the year by solving this puzzle.

- My units digit is equal to the positive root of the equation $x^2 + 4x - 21 = 0$.

- Find two numbers whose sum is 24 and product is 108. The two-digit number formed by my thousands and hundreds digit is the larger of the two numbers. My tens digit is the smaller of the two numbers.

  - The sum of my digits is equal to x: $\sqrt{\frac{1}{2}x} + 4 = 7$.

    What year am I?

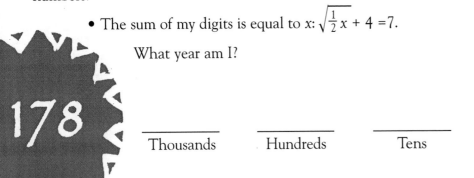

178

Thousands	Hundreds	Tens	Units

On December 1 of this year, Rosa Parks was arrested in Montgomery, Alabama, for refusing to give up her seat and move to the back of the bus. Her arrest triggered a yearlong boycott of the city bus system and helped end racial segregation on municipal buses in the South. This event has been called the beginning of the civil rights movement in the United States.

- If $f(x) = 2x^2$ and $g(x) = x + 5$, then the two-digit number formed by my tens and units digits is equal to $g(f(x))$ when $x = 5$.

- My hundreds digit is equal to x: $\sqrt{4x} = 6$.

- The sum of my digits is equal to $_5P_2$.

What year am I?

| _____ | _____ | _____ | _____ |
| Thousands | Hundreds | Tens | Units |

179

On December 15 of this year, the Bill of Rights, the first ten amendments to our Constitution, became effective following the approval of Virginia. Find the year of this historical highlight by solving this puzzle.

- My units digit is the absolute value of the product:
$$\frac{m2+2m+1}{m^2-1} \times \frac{1-m}{1+m}$$

- My hundreds digit is equal to the x-coordinate and my tens digit is equal to the y-coordinate of the vertex of the parabola $-(x-7)^2 + 9$.

- The sum of my digits is equal to x: $\sqrt{2x} = 6$.

What year am I?

180

_____   _____   _____   _____
Thousands   Hundreds   Tens   Units

States of the Union
1. Delaware, Pennsylvania, and New Jersey became states in 1787.
2. Georgia, Massachusetts, Connecticut, New Hampshire, South Carolina, Virginia, New York, and Maryland all became states in 1788.
3. North Carolina became a state in 1789.
4. Rhode Island became a state in 1790.
5. Vermont became a state in 1791.
6. Kentucky became a state in 1792.
7. Tennessee became a state in 1796.
8. Ohio became a state in 1803.
9. Louisiana became a state in 1812.
10. Indiana became a state in 1816.
11. Mississippi became a state in 1817.
12. Illinois became a state in 1818.
13. Alabama became a state in 1819.
14. Maine became a state in 1820.
15. Missouri became a state in 1821.
16. Arkansas became a state in 1836.
17. Michigan became a state in 1837.
18. Florida and Texas became states in 1845.
19. Iowa became a state in 1846.
20. Wisconsin became a state in 1848.
21. California became a state in 1850.
22. Minnesota became a state in 1858.
23. Oregon became a state in 1859.
24. Kansas became a state in 1861.
25. West Virginia became a state in 1863.
26. Nevada became a state in 1864.
27. Nebraska became a state in 1867.
28. Colorado became a state in 1876.
29. Montana, North Dakota, South Dakota, and Washington became states in 1889.
30. Idaho and Wyoming became states in 1890.
31. Utah became a state in 1896.
32. Oklahoma became a state in 1907.
33. New Mexico and Arizona became states in 1912.
34. Alaska and Hawaii became states in 1959.

Famous Firsts
35. The first junior high school opened in 1910.
36. Nellie Tayloe Ross became the first woman governor in 1925.
37. The Panama Canal opened in 1914.

Daily Warm-Ups: Algebra II

38. Hattie Caraway became the first woman U.S. Senator in 1932.
39. Robert C. Weaver became a member of Johnson's cabinet in 1966.
40. Driver's licenses were required in 1937.
41. Dr. Blackwell received her degree in 1849.
42. Justice Brandeis was appointed in 1916.
43. *These Are My Children* aired in 1949.
44. Maggie Walker became a bank manager in 1903.
45. The first "singing telegram" was delivered in 1933.
46. *American Magazine* was published in 1747.
47. Glenn orbited Earth in 1962.
48. The school for the blind was opened in 1829.
49. The first auto was driven in 1896.
50. The parachute jump was made in 1912.
51. *Freedom's Journal* was first published in 1827.
52. The swallows first returned in 1776.
53. The first women's basketball game was played in 1893.
54. Hank Aaron hit the home run in 1974.
55. Robinson was recruited in 1947.
56. The first shot at the Battle of Lexington was fired in 1775.
57. The first Seeing Eye dog was presented in 1923.
58. Gwendolyn Brooks won the Pulitzer Prize in 1950.
59. de Soto reached the Mississippi River in 1541.
60. The railroad was completed in 1869.
61. The planetarium was opened in 1930.
62. Sally Ride's mission was in 1983.
63. Neil Armstrong landed on the moon in 1969.
64. The balloon crossing occurred in 1978.
65. The baby was born in 1893.
66. The newspaper was published in 1690.
67. Thurgood Marshall became a justice in 1967.
68. The women became FBI agents in 1972.

Discoveries, Inventions, and Notable Accomplishments
69. The X-ray was discovered in 1895.
70. "The Landlord's Game" was patented in 1904.
71. The accordion was patented in 1854.
72. The Pentagon was completed in 1943.
73. Gold was discovered in 1848.
74. The phone call was made in 1877.

75. The tomb was opened in 1923.
76. The phonograph was invented in 1878.
77. The electric razor was patented in 1931.
78. Paper money became legal tender in 1862.
79. Jonas Salk introduced the polio vaccine in 1953.
80. Teflon® was invented in 1938.
81. Radium was isolated in 1902.
82. The Barbie® doll was invented in 1959.
83. The disposal diaper was invented in 1950.
84. Play-Doh was invented in 1956.
85. Sarah Boone invented her ironing board in 1892.
86. The Slinky® was invented in 1945.
87. The Simplon Tunnel was opened in 1906.
88. Lindbergh crossed the Atlantic in 1927.
89. The Brooklyn Bridge was opened in 1883.
90. Liquid paper was invented in 1956.
91. The Dionne quintuplets were born in 1934.
92. "Casey at the Bat" was printed in 1888.
93. The vacuum cleaner was patented in 1869.
94. Benjamin Franklin conducted his experiment in 1752.
95. The wheat reaper was patented in 1834.

96. Patsy Sherman accidentally discovered Scotchgard™ in 1952.
97. The bar code was patented in 1952.
98. The Holland Tunnel was opened in New York in 1927.
99. The *Spruce Goose* made its flight in 1947.
100. Evaporated milk was patented in 1884.
101. The painting was found in 1961.
102. The Louisiana Purchase took place in 1803.

Happy Birthday to You

103. Paul Revere was born in 1735.
104. J.R.R. Tolkien was born in 1892.
105. Dr. Martin Luther King, Jr., was born in 1929.
106. A.A. Milne was born in 1882.
107. Maria Tallchief was born in 1925.
108. Mozart was born in 1756.
109. Hank Aaron was born in 1934.
110. Wilder was born in 1867.
111. Grant Wood was born in 1892.
112. Galileo was born in 1564.
113. Susan B. Anthony was born in 1820.
114. Shalom Aleichem was born in 1859.

115. Copernicus was born in 1473.
116. Renoir was born in 1841.
117. Dr. Seuss was born in 1904.
118. Michelangelo was born in 1475.
119. Burbank was born in 1849.
120. Einstein was born in 1879.
121. Juárez was born in 1806.
122. Emmy Noether was born in 1884.
123. Haydn was born in 1732.
124. Jagjivan Ram was born in 1908.
125. Booker T. Washington was born in 1856.
126. Frances Perkins was born in 1882.
127. Anne Sullivan was born in 1866.
128. Euler was born in 1707.
129. Shakespeare was born in 1564.
130. Richter was born in 1900.
131. Gauss was born in 1777.
132. Edward Lear was born in 1812.
133. Nightingale was born in 1820.
134. Frank Baum was born in 1856.
135. Maria Agnesi was born in 1718.
136. Edward Jenner was born in 1749.

137. Malcolm X was born in 1925.
138. Mesmer was born in 1734.
139. Socrates was born in 469 B.C.E.
140. Anne Frank was born in 1929.
141. Lazarus was born in 1849.
142. Jim Henson was born in 1936.
143. Chubby Checker was born in 1941.
144. Annie Peck was born in 1850.
145. Bill Mauldin was born in 1921.
146. Edmund Halley was born in 1656.
147. Möebius was born in 1790.

Historical Highlights
148. The first election was held in 1789.
149. The League of Nations was established in 1920.
150. The schools closed in 1982.
151. The record snowfall occurred in 1881.
152. *Challenger* exploded in 1986.
153. *The Lone Ranger* was heard in 1933.
154. Supreme Court first met in 1790.
155. Japan was founded in 660 B.C.E.
156. The Marines planted the flag in 1945.
157. The siege of the Alamo began in 1836.

158. The last episode of M*A*S*H appeared in 1983.
159. "The Star-Spangled Banner" became our national anthem in 1931.
160. Harriet Tubman died in 1913.
161. The Salvation Army was established in 1880.
162. Earth Day was founded in 1979.
163. Patrick Henry gave his speech in 1775.
164. The A.S.P.C.A. was founded in 1866.
165. The big wind occurred in 1934.
166. The 335-year war ended in 1986.
167. The *Hindenburg* exploded in 1937.
168. The *Lusitania* was sunk in 1915.
169. Native Americans were granted citizenship in 1924.

170. The Magna Carta was signed in 1215.
171. "Wrong Way" Corrigan made the flight in 1938.
172. The English colony was founded in 1583.
173. Pictures of Earth were seen in 1959.
174. The planet Neptune was observed in 1846.
175. The Statue of Liberty was given to the United States in 1886.
176. The blackout occurred in 1965.
177. Mickey Mouse debuted in 1928.
178. The Gettysburg Address was delivered in 1863.
179. Rosa Parks took her stand in 1955.
180. The Bill of Rights became effective in 1791.

combinations of *n* elements taken *m* at a time—$\binom{n}{m}$ or $_nC_m = \dfrac{n!}{(n-m)!m!}$.

determinant—A real number associated with each square matrix. For example in the matrix $\begin{bmatrix} a & b \\ c & d \end{bmatrix}$, the determinant is $ad - bc$.

diagonal method—A method used to evaluate the determinant of a 3×3 matrix, where the first and second column become the fourth and fifth column. The determinant is found by adding the products of the three diagonals going from upper right to lower left and then subtracting the three diagonals going from lower left to upper right.

discriminant—The expression $b^2 - 4ac$ is for a quadratic equation $ax^2 + bx + c = 0$.

distance formula—The distance between any two points (x_1, y_1) and (x_2, y_2) is given by the formula $d = \sqrt{(x^2 - x^1)^2 + (y^2 - y^1)^2}$.

geometric mean—The geometric mean of *a* and *b* is $\sqrt{ab}$.

logarithm—If $a > 0$ and $x > 0$, $y = \log_a x$ if and only if $x = a^x$.

parabola—The graph of a quadratic equation.

permutation—An arrangement of objects where order is important: $_nP_m = \dfrac{n!}{(n-m)!}$.

quadratic equation—An equation that can be written in the form $ax^2 + bx + c = 0$, where $x \neq 0$.

quadratic formula—$x = \dfrac{-b \pm \sqrt{b^2 - 4ac}}{2a}$, where *a*, *b*, and *c* are real numbers and $a \neq 0$.

radian—A unit of angle, arc, or rotation measure such that π radians equal $180°$ or $\dfrac{\pi}{180} = 1$.

Daily Warm-Ups: Algebra II

slope of a line—The ratio of the vertical change to the horizontal change; $\dfrac{y_2 - y_1}{y_2 - y_1}$.

solution of a system—An ordered pair that is a solution of each of the equations in the system.

solution of a system of equations in x, y, and z—An ordered triple (x, y, z) that is a solution in each of the equations in the system.

triangular numbers—A figurate number that is defined by the equation $\dfrac{n(n + 1)}{2}$. The first four triangular numbers:

vertex of a parabola—The lowest point of a parabola that opens up and the highest point of a parabola that opens down.

***x*-intercept**—The value of x when $y = 0$.

***y*-intercept**—The value of y when $x = 0$.

Turn downtime into learning time!

Other books in the

Daily *Warm-Ups* series:

- Algebra
- Analogies
- Biology
- Character Education
- Chemistry
- Commonly Confused Words
- Critical Thinking
- Earth Science
- Geography
- Geometry
- Journal Writing
- Mythology
- Physics

- Poetry
- Pre-Algebra
- Prefixes, Suffixes, & Roots
- Shakespeare
- Spelling & Grammar
- Test-Prep Words
- U.S. History
- Vocabulary
- World Cultures
- World History
- World Religions
- Writing